DISSECTION MANUAL

COMPANION TO ROHEN/YOKOCHI COLOR ATLAS OF ANATOMY

SECOND EDITION

DISSECTION MANUAL

COMPANION TO ROHEN/YOKOCHI COLOR ATLAS OF ANATOMY

SECOND EDITION

Jack L. Wilson, Ph.D.

Associate Professor
Department of Anatomy and Neurobiology
The University of Tennessee, Memphis
The Health Science Center
Memphis, Tennessee

IGAKU-SHOIN MEDICAL PUBLISHERS, INC.
New York • Tokyo

Interior Design by Wanda Lubelska Design
Cover Design by Paul Agule Design
Typesetting by Achorn Graphic Services, Inc.
Printing and Binding by Arcata Graphics/Kingsport

Published and distributed by

IGAKU-SHOIN Medical Publishers, Inc.
1140 Avenue of the Americas, New York, N.Y. 10036

IGAKU-SHOIN Ltd.,
5-24-3 Hongo, Bunkyo-ku, Tokyo

Wilson, Jack L.
    Dissection manual : companion to Rohen/Yokochi Color atlas of
anatomy, 2nd edition / Jack L. Wilson. — 2nd ed.
        p.   cm.
    Includes index.
    1. Anatomy, Human—Laboratory manuals.   2. Human dissection—
Laboratory manuals.   I. Rohen, Johannes W. (Johannes Wilhelm).
Color atlas of anatomy. II. Title.
    [DNLM: 1. Anatomy—laboratory manuals.   2. Dissection—laboratory
manuals.   QS 17 R737c Suppl.]
    QM25.R55   1988 Suppl.
    611—dc19
    DNLM/DLC
    for Library of Congress   87-33862
                                                                                        CIP

ISBN: 0-89640-142-1 (New York)
ISBN: 4-260-14142-2 (Tokyo)

Printed and bound in USA

10 9 8 7 6 5 4 3 2 1

PREFACE

It is my hope that this dissector will be a useful aid for students studying gross anatomy. This text offers an efficient approach to dissection tailored to current curricula and supplies the basic technical descriptions that students need for carrying out dissections in a logical sequence. Additional information is included that describes many relationships observed during dissection as well as other primary features of the areas. However, this dissector is not intended to be a substitute for a textbook or atlas because it does not repeat the many illustrations and detailed information available in such books. It provides the basic descriptions and illustrations that are required for performing dissections. Thus, an atlas and a textbook must be used along with this dissector, and all three books need to be available and used in the laboratory.

This second edition of *Dissection Manual* incorporates many suggestions and comments received from readers. Many new illustrations have been added to improve the effectiveness and usefulness of the manual. Parts of the text have been rewritten to provide better descriptions of the dissection instructions. It is hoped that this book will meet the needs of students in an efficient, new and innovative manner.

This dissector is divided into regional chapters with sequentially numbered units, making it easily adaptable to virtually any course sequence or atlas. The text descriptions are correlated by page numbers (in parentheses) with the *Color Atlas of Anatomy* (Rohen and Yokochi, Igaku-Shoin Medical Publishers, 1988). However, it should be noted that this dissector can also be easily used with any other currently available atlas.

I would like to express my deep appreciation to a number of friends and colleagues with whom I have been privileged to work. Special thanks is due Dr. Clark E. Corliss for his encouragement and excellent editorial assistance during preparation of the manuscript. I was fortunate to have had the opportunity to learn a great deal from such distinguished model teachers and anatomists as Drs. Sidney A. Cohn, Alexander A. Fedinec, Jean R. Holbrook and Harry H. Wilcox. I am also grateful to Prof. Rohen and his staff of artists, particularly Mrs. Christiane Wittek and Mr. Anton Atzenhofer, for their help in preparation of the illustrations. To the staff of Igaku-Shoin, I express my appreciation for their support and assistance during the development of this book. Special thanks are due to Ms. Mary D. Gaither and Ms. Margaret Bush for their excellent typing and patience during preparation of the manuscript.

I am most grateful to my wife, Rose Marie, and my children, Lynn, John and Marilyn, all of whom have been so loving and supportive. Finally, I give a very special thanks to my students, those in the past and those yet to come, who make my profession such a pleasant and enriching experience.

JACK L. WILSON

Memphis, Tennessee

CONTENTS

CHAPTER ONE
INTRODUCTION   1

Care of Cadaver, 2
Dissection Techniques, 2

CHAPTER TWO
BACK AND POSTERIOR NECK   7

Surface Anatomy, 8
Skin Reflection, 8
Cutaneous Nerves, 9
Muscles of the Back, 10
    Superficial Group, 10
    Middle Group, 12
    Deep Group, 12
Suboccipital Triangle, 15
Vertebral Column and Spinal Cord, 16

CHAPTER THREE
UPPER LIMB   19

Review of the Superficial Muscles of the Back, 20
Posterior Shoulder and Deltoid Regions, 20
Cutaneous Vessels and Nerves, 23
Pectoral Region, 26

Axilla, 28
    Axillary Vein, 29
    Axillary Artery, 29
    Brachial Plexus, 31
    Axillary Lymph Nodes, 33
Arm, 33
    Anterior (Flexor) Compartment, 34
    Posterior (Extensor) Compartment, 37
Cubital Fossa, 40
Forearm, 41
Extensor Compartment of the Forearm and Dorsum of the Hand, 42
    Superficial Muscle Group, 42
    Deep Muscle Group, 43
    Vessels and Nerves, 44
    Dorsum of the Hand, 44
Flexor Compartment of the Forearm, 44
    Superficial Muscle Group, 45
    Deep Muscle Group, 45
    Vessels and Nerves, 45
    Deep Dissection, 46
Wrist (Ventral Surface), 49
Palm of the Hand, 51
    Thenar Compartment, 51
    Hypothenar Compartment, 52
    Central Compartment, 52
    Interosseous-Adductor Compartment, 54
Articulations, 55
    Shoulder Joint, 55
    Elbow, 56
    Distal Radioulnar and Wrist Joints, 56

CHAPTER FOUR
LOWER LIMB   57

Superficial Structures, 58
Gluteal Region, 58
    Muscles, 59
    Vessels and Nerves, 61
Anterior (Extensor) Compartment of the Thigh, 62

Muscles, 67

Nerves, 68

Vessels, 69

Medial (Adductor) Compartment of the Thigh, 70

Posterior (Flexor) Compartment of the Thigh, 71

Popliteal Fossa, 72

Leg, 73

Anterior Compartment of the Leg and Dorsum of the Foot, 74

Muscles, 74

Vessels and Nerves, 76

Lateral Compartment of the Leg, 77

Posterior Compartment of the Leg, 77

Superficial Compartment, 78

Deep Compartment, 79

Plantar Surface of the Foot, 81

First Layer, 82

Second Layer, 85

Third Layer, 85

Fourth Layer, 86

Ligaments of the Plantar Surface of the Foot, 87

Plantar Calcaneonavicular (Spring) Ligament, 87

Long Plantar Ligament, 87

Plantar Calcaneocuboid Ligament, 87

Articulations, 87

Hip Joint, 88

Knee Joint, 88

Ankle Joint, 89

Joints of the Foot, 89

CHAPTER FIVE
THORAX

91

Surface Anatomy, 92

Thoracic Wall, 92

Thoracic Cavity and Pleura, 96

Superior Mediastinum, 98

Middle Mediastinum and Pericardium, 100

External Features of the Heart, 102

Coronary Arteries, 103

Internal Features of the Heart, 104
Right Atrium, 104
Right Ventricle, 105
Left Atrium, 106
Left Ventricle, 106
Conducting System of the Heart, 107
Pulmonary Vessels and Main Bronchi, 108
Lungs, 109
Bronchopulmonary Segments, 110
Review of Mediastinal Relationships, 110
Posterior Mediastinum, 110

CHAPTER SIX
ABDOMEN   115

Superficial Structures and Landmarks, 116
Cutaneous Nerves and Blood Vessels, 117
Anterolateral Muscles and Fasciae, 118
External Abdominal Oblique Muscle, 119
Internal Abdominal Oblique Muscle, 119
Transversus Abdominis Muscle, 121
Rectus Abdominis Muscle and Sheath, 122
Inguinal Region and Canal, 123
Spermatic Cord, 127
Peritoneum, 127
Small and Large Intestines, 130
Stomach and Spleen, 134
Celiac Trunk, 136
Splenic Artery, 136
Left Gastric Artery, 137
Common Hepatic Artery, 137
Duodenum and Pancreas, 138
Liver and Gallbladder, 140
Kidneys and Suprarenal Glands, 142
Diaphragm, 144
Posterior Abdominal Wall, 145
Muscles, 145
Vessels, 145

Lumbar Plexus of Nerves, 146
Autonomic Nervous System, 147

CHAPTER SEVEN
PERINEUM   153

Boundaries and Landmarks, 154
Anal Triangle, 154
Urogenital Triangle, 157
    Male, 157
    Female, 162

CHAPTER EIGHT
PELVIS   167

Boundaries and Landmarks, 168
Bony Pelvis, 168
Peritoneum, 168
Rectum, 170
Midsagittal Section, 170
Ureter and Urinary Bladder, 172
Pelvic Viscera, 174
    Male, 174
    Female, 175
Pelvic Floor and Walls, 177
    Muscles and Fasciae, 177
    Nerves, 179
    Vessels, 180

CHAPTER NINE
HEAD AND NECK   185

Surface Anatomy of the Neck, 186
Skin Reflections and Superficial Structures of the Neck, 186
Anterior Triangle, 190
    Muscular Triangle, 190
    Carotid Triangle, 192

Posterior Triangle, 195
Scalp, 202
Cranial Cavity, 202
    Bony Landmarks, 202
    Meninges, 204
    Dural Venous Sinuses, 204
Pharynx, 205
    Disarticulation of the Head, 206
    Exterior of the Pharynx, 207
    Interior of the Pharynx, 211
Larynx, 211
Submandibular Region, 214
Face, 218
    Muscles of Facial Expression, 218
    Facial Nerve and Vessels, 220
Parotid Region, 221
    Bisection of the Head, 221
    Parotid Gland, 222
Temporal and Infratemporal Fossae, 224
Deep Dissection of the Infratemporal Fossa, 230
Nasal Cavity, 231
Oral Cavity, 233
Palate and Pharyngeal Cavity, 234
Pterygopalatine Fossa, 236
Orbital Region, 238
    Eyelid and Lacrimal Apparatus, 238
    Orbit, 238
Ear, 242
    Inner Ear, 242
    Middle Ear, 243
    External Ear, 244

CHAPTER ONE

INTRODUCTION

1. CARE OF CADAVER

As a student of gross anatomy, you have the unique honor and privilege of dissecting the human body. This opportunity has been provided mostly by persons who have willed their bodies for your learning experience. However, with this privilege, there is the continuing responsibility to treat cadavers with respect and dignity.

The cadaver requires care on a regular basis to ensure the most effective learning experience from dissection of the body. A major problem is the possibility that the cadaver may dry out. Therefore, at the end of each dissection period, you should moisten the body with the wetting solution provided and carefully wrap the cadaver. If the cadaver is kept in a metal table, always keep the tabletop closed when the body is not in use. Proper preservation of the body provides many months of rewarding laboratory work.

2. DISSECTION TECHNIQUES

Essential to an effective study and dissection of the human body is an adequate set of dissecting instruments. The basic instruments required are shown in Figure 1.1. The instruments should be made of good-quality steel, not of plastic or any other material that might easily break and be dangerous. A great deal of your work can be done best by blunt dissection with a probe, forceps and scissors. Scissors are especially helpful for separating structures when the blades are inserted into the connective tissue and the handles opened and closed to follow fascial planes. A scalpel is used primarily for cutting and reflecting skin, muscles and other large structures. In general, a scalpel is not used for blunt or detailed dissection because of the risk of damage to many structures. Dissection techniques are learned with practice, and patience is needed to develop skill over time. Dissection efforts should be shared by *all* students working together at a dissection table.

The dissections described in this book are based on a regional approach. As dissection of a new region is begun, time should first be devoted to observation and palpation of surface landmarks mentioned in this manual and your textbook. It is also important to make a complete study of the bones of each region as you begin to dissect. These aspects of dissection are essential for a comprehensive understanding of gross anatomy.

Your first responsibility in each region will be removing the skin and superficial fascia (tela subcutanea). General guidelines that describe where skin flaps should be made are provided in the text. It is important that the skin and superficial fascia be removed in *two*

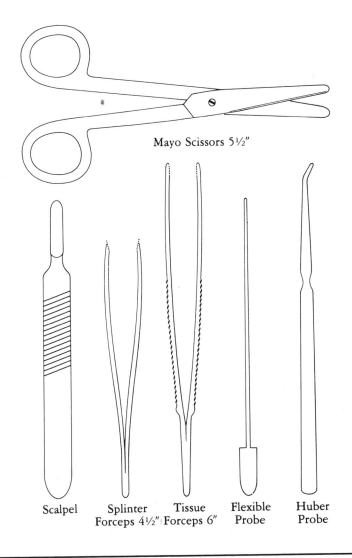

Mayo Scissors 5½"

Scalpel Splinter Tissue Flexible Huber
Forceps 4½" Forceps 6" Probe Probe

Figure 1.1

stages instead of collectively in one operation. The plane on the deep surface of the skin can be identified by the pitted appearance on its undersurface. To remove the skin at this plane, make superficial cuts in the skin along the lines of reflection illustrated in the dissector. Put tension on the corner of the skin flaps and, with the scalpel *parallel* to the surface of the body, separate the skin from the superficial fascia at the plane noted previously (Figure 1.2). The exposed superficial fascia contains variable amounts of fat and connective tissue. Most of the fascia can be removed by blunt dissection. However, in areas where the fascia is more dense, it may be removed with

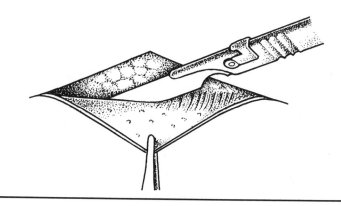

Figure 1.2

the scalpel, again held parallel to the body. Through this superficial fascia, numerous cutaneous blood vessels and nerves reach the skin. Some of these blood vessels and nerves will be mentioned in the dissector, but it is not necessary to dissect and identify each one in detail. After removing the superficial fascia, you will study and identify the deep muscles, vessels, nerves and fascial layers. The names of the major anatomic structures appear in this dissector in **bold** type. You should familiarize yourself with these terms. Always clearly dissect and identify the borders of the muscles, removing enough deep fascia to determine the direction of the muscle fibers. When developing a neurovascular bundle, be sure to dissect and separate the nerves and vessels in detail, following them *completely* to their origins and terminations. It is essential that you complete each stage of the work *before* proceeding to the next part described in this dissector. After an area has been completely dissected, it is beneficial for continuity to review in detail the dissection and relate it to previous dissections. This approach will allow you to appreciate the body as a whole, rather than as isolated regions.

Each dissection team should always have at their table complete sets of textbooks, dissectors, atlases and dissection instruments. References to a textbook and an atlas appear in this dissector, and it is difficult to work effectively without using these books. Thus, a textbook and an atlas must be used concurrently with the dissector in the laboratory. This dissector will guide your daily sequence of dissection; your textbook will provide details of the structures and their relationships; and your atlas will provide a visual model for your dissection, allowing you to identify and observe relationships of the exposed structures. Effective learning of gross anatomy is based on active experience; it is something you must do. Comprehension of basic concepts and visualization are your best learning methods—

not rote memorization. Therefore, you should direct your learning toward these ends. In the laboratory, every student must share the responsibilities of reading from the textbook and dissector and carrying out dissections. Learn to be independent and thorough in your work. Always appreciate variations from normal anatomy, which you will find occur with great frequency.

Before beginning this course, familiarize yourself with the terms of position, movement and relationships explained in the introductory chapter of your textbook. Keep in mind that all descriptions in this dissector are based on the anatomic position.

CHAPTER TWO

BACK AND POSTERIOR NECK

3. SURFACE ANATOMY

On the posterior aspect of the occipital bone of the skull, identify the **external occipital protuberance** (p. 33). Descending from this point in the midline of the neck and the back is the vertebral furrow. The first vertebral spine palpated in this furrow is that of the seventh cervical vertebra, the **vertebra prominens** (p. 174). The furrow is deepest in the thoracic and lumbar regions, where all spines of the thoracic and lumbar vertebrae can be felt. Inferiorly, the furrow ends at the **sacrum,** from which the **iliac crest** (p. 401) can be followed laterally until it reaches its highest point at the level of the fourth lumbar vertebra.

At the tip of the shoulder, identify the **acromion** of the scapula. The acromion is the most lateral point of the shoulder and articulates anteriorly with the lateral end of the **clavicle.** The acromion continues medially as the **spine of the scapula** (p. 344), which leads to the **medial (vertebral) border** of the scapula. The scapular spine lies at the level of the third thoracic vertebral spine. The **inferior angle of the scapula,** located at the level of the seventh thoracic vertebral spine, can be felt at the inferior end of the medial border (p. 343). The bulge of the **erector spinae** muscle can be identified in the lumbar region on either side of the furrow. Also, the **posterior axillary fold,** formed by the lower border of the **latissimus dorsi** and **teres major** muscles, can be felt at the posterior shoulder.

4. SKIN REFLECTION

The skin and superficial fascia (tela subcutanea) should be removed in *two* stages. Initially, remove the skin as described in the text that follows; then remove the tela subcutanea. Follow the incision lines shown in Figure 2.1. With the scalpel, make a vertical incision from A to G, being careful to cut only through the skin and not deeply into the superficial fascia. Transverse skin incisions should then be made as indicated. The **A** to **A** transverse incisions should be carried out as far as the mastoid process; the **B** to **B** incisions across the top of the shoulder should be carried out as far as the acromion; and the other cuts (**C, D, E** and **F**) should be carried out laterally to the *midaxillary line.* It is important that the skin flaps be taken around the sides of the body to the midaxillary line. The plane between the skin and the superficial fascia can be identified by the pitted appearance of the undersurface of the skin. As the superficial fascia is removed, watch for examples of the cutaneous nerves and vessels discussed in the text that follows.

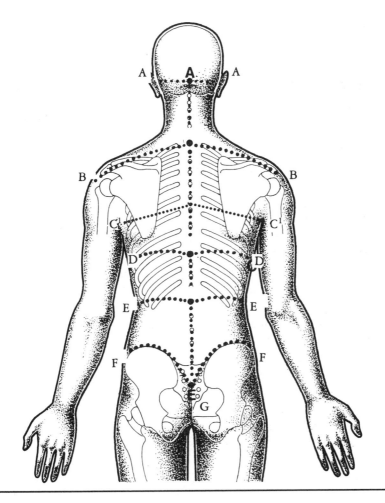

Figure 2.1

5. CUTANEOUS NERVES

The cutaneous nerves to the skin of the back and posterior neck are branches of the dorsal rami of the spinal nerves (pp. 204–206; 212–213). These nerves are segmental and are derived from most of the 31 pairs of spinal nerves. Read the description of the cutaneous nerves of the back in your textbook and understand their general arrangements. It is necessary only to identify a few of these cutaneous nerves on each side of the body. They traverse the deep muscles to enter the skin and can be found by teasing them from the superficial fascia. It is often difficult to distinguish the nerves from the fascia, and they are sometimes more easily seen during later reflections of the deep muscles of the back. Save the nerves you identify so that they can be followed to the spinal cord in a later dissection.

6. MUSCLES OF THE BACK

The muscles of the back are divided into a superficial group that functionally belongs to the upper limb, a middle group that is respiratory in function and a deep group that is composed of the true (intrinsic) muscles that move the vertebral column.

A. Superficial Group

The superficial group is composd of the **trapezius, latissimus dorsi, rhomboideus major, rhomboideus minor** and **levator scapulae** muscles (pp. 200; 354–355). These muscles originate from the vertebral column and insert on the scapula or humerus (Figure 2.2). The details of their attachments and innervations should be studied in your textbook as you dissect. Describe the components and functions of the **pectoral girdle.** The trapezius and latissimus dorsi form a superficial layer and are extensive, covering most of the back. Carefully remove the tela subcutanea from the surface of these two muscles with blunt dissection and the scalpel. Expose their borders and note the directions of the fibers.

The **trapezius** has an extensive origin at the midline, extending from the skull to the twelfth thoracic vertebral spine and inserting on the clavicle, acromion and scapular spine. After cleaning the surface of this muscle, use scissors to begin reflecting it inferiorly 2.5 cm from each side of the twelfth thoracic spine and continue the cut superiorly to the posterior neck (Figure 2.2, A-B). Use blunt dissection as you proceed superiorly to separate the trapezius from the underlying muscles, which will be studied next. The upper portion of the trapezius on the neck is more fragile and care should be taken not to tear these thin muscle fibers. Do not dissect into the lateral side of the neck beyond the lateral border of the trapezius. After cutting the muscle, carry the trapezius laterally and cut its insertion to the **scapular spine** and **acromion** (Figure 2.2, C-D), leaving the clavicular fibers intact. Deep to the reflected portion of the trapezius, carefully dissect and identify branches of the **transverse cervical artery** and **vein** and **subtrapezial nerve plexus** (formed by the accessory nerve [p. 214] and third and fourth cervical spine nerves) from the connective tissue on the deep surface of the muscle (Figure 2.2). With the probe and scissors, carefully remove the fascia in this area to identify these vessels and nerves.

The position and relationship of the **latissimus dorsi** should be

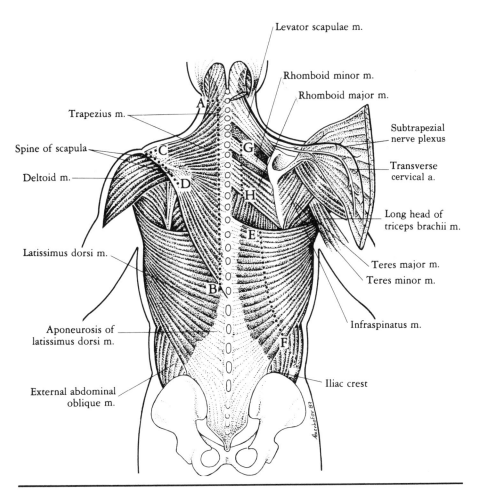

Levator scapulae m.

Rhomboid minor m.

Rhomboid major m.

Trapezius m.

Subtrapezial nerve plexus

Spine of scapula

Transverse cervical a.

Deltoid m.

Long head of triceps brachii m.

Latissimus dorsi m.

Teres major m.

Teres minor m.

Aponeurosis of latissimus dorsi m.

Infraspinatus m.

External abdominal oblique m.

Iliac crest

Figure 2.2

reviewed before reflection. This muscle can be extremely thin, especially in its superior part. Note that medially, the superior border is deep to the trapezius and crosses the inferior angle of the scapula laterally. Inferiorly and laterally, the fibers of this muscle interdigitate with those of the external abdominal oblique muscle and attach to the iliac crest. Cut transversely through the fibers of the latissimus dorsi and reflect the muscle laterally toward the axilla (Figure 2.2, E-F). Blunt dissect deep to the aponeurosis of the muscle and separate the latissimus dorsi from the **posterior inferior serratus muscle** (Figure 2.4), which may easily adhere to the underside of the latissimus (p. 205). Note that superiorly, the fibers of the latissimus converge to wrap around the inferior border of the **teres major** muscle, where they form the **posterior axillary fold** at the shoulder. Reflect the latissimus toward the arm and separate and identify the teres major (Figure 2.2). Also, during reflection, carefully identify and separate the fascial plane deep to the latissimus dorsi at the posterior fold to observe the **serratus anterior** muscle. The innervation and blood supply (thoracodorsal nerve and vessels) and the insertion (intertubercular groove of the humerus) of the latissimus will be more easily seen during dissection of the axilla and arm.

Deep to the reflected trapezius, identify the **rhomboideus major** and **minor** and **levator scapulae** muscles (pp. 200; 355). Review their origins, insertions and innervations. These muscles attach the scapula to the vertebral column and are important in scapular movement. Reflect the two rhomboid muscles at their midpoints (Figure 2.2, G-H), leaving the levator scapulae intact to maintain the position of the scapula. Protect the **posterior superior serratus** muscle (described in the text that follows) deep to the rhomboid muscles as they are reflected. The **dorsal scapular vessels** and **nerve** supply the two rhomboids and levator scapulae muscles (the levator is also innervated by the third and fourth cervical spinal nerves). Identify the dorsal scapular nerve and vessels as they pass from the deep surface of the levator scapulae into the fascia on the deep aspect of the rhomboid muscles along their insertions on the vertebral border of the scapula (Figure 2.3).

B. Middle Group

The middle group of back muscles consists of two respiratory muscles (Figure 2.4). The first muscle is the **posterior superior serratus** (p. 205), which is deep to the rhomboid muscles. The second muscle is the **posterior inferior serratus** (p. 205), which is deep to the latissimus dorsi. These two muscles are extremely thin and should be reflected by cutting through the middle of their fibers. These muscles are innervated by the intercostal nerves, located deep to each muscle.

C. Deep Group

The deepest muscles of the back are referred to as the intrinsic (true) muscles because they function in direct movements of the vertebral column and skull. As a group, these muscles have a complex arrangement, the details of which are beyond the requirements of most courses. Therefore, the dissections that follow are organized so that any subsequent section can be omitted. All these muscles receive their motor innervation from the dorsal rami of the spinal nerves, unlike the superficial and middle groups, which receive innervation from the ventral rami.

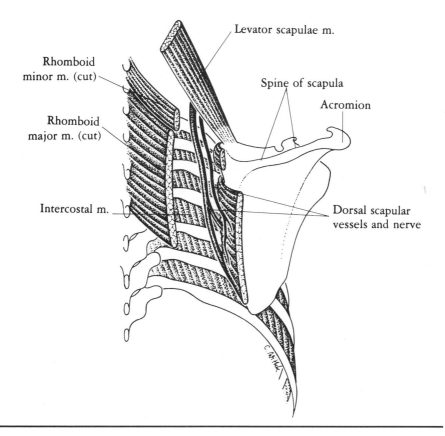

Figure 2.3

Splenius Muscle. This muscle is on the posterior aspect of the neck and appears as a bandage (Figure 2.4), aiding support of the deep muscles of the neck (pp. 200; 212–213). It is immediately deep to the superior fibers of the trapezius, which have to be completely reflected to expose the splenius muscle and other deep structures of the neck. The splenius has capitis and cervicis parts that overlie the deep erector spinae and semispinalis muscles. Identify these two parts of the splenius muscle, cut their origins from the cervical and upper thoracic vertebral spines, and reflect the muscle laterally.

Erector Spinae Muscle. This muscle is the large, vertical, deep muscle mass that forms the prominent bulge on both sides of the lumbar vertebrae (p. 201). The erector spinae is enclosed in a dense layer of deep fascia, the **thoracolumbar fascia** (p. 200), which has two laminae attached to the spinous processes (posterior layer) and transverse processes (anterior layer) of the vertebrae. The posterior layer of the thoracolumbar fascia should be incised vertically on each

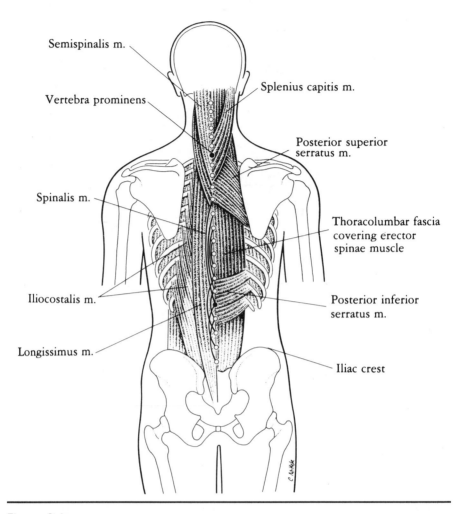

Figure 2.4

side of the vertebral furrow from the sacrum to the base of the neck.
You should then identify and separate the three columns (Figure 2.4)
of the erector spinae: the **iliocostalis** (lateral column), **longissimus**
(middle column) and **spinalis** (medial column). The fibers of the iliocostalis insert primarily at the angles of the ribs, with some fibers
reaching as high as the fourth cervical vertebra. The intermediate
fibers of the longissimus insert at the levels of the thoracic and cervical vertebrae, with the highest fibers reaching the mastoid process of
the skull. The most medial part of the erector spinae is the spinalis,
which spans the spinous processes. It is most prominent in the thoracic region. If they were not observed previously, identify the cutaneous nerves of the back (dorsal rami), which pass between the longissimus and iliocostalis columns.

Transversospinalis Group. This group of muscles is so named because they have their origins from the vertebral transverse processes and insert into the spinous processes. Included in this group are the semispinalis, multifidus and rotatores. The **semispinalis** (Figure 2.4) covers the upper half of the length of the vertebral column and forms thoracic, cervical and capitis portions (pp. 201–203; 214–215). The most prominent part is the semispinalis capitis, which is immediately deep to the splenius capitis and can be identified after reflection of this muscle. The semispinalis is the largest muscle mass at the posterior neck. Identify the **greater occipital nerve** as it passes through this muscle (pp. 212–216). The multifidus and rotatores are the deepest muscles of the transversospinalis group and are prominent in the lower thoracic and lumbar regions. They lie deep to the spinalis and extend most of the length of the vertebral column. The details of their attachments are complex and, if you are so instructed, should be studied in your textbook and atlas.

7. SUBOCCIPITAL TRIANGLE

This triangle is the deepest area of the posterior neck and its dissection provides access to the superior courses of the **vertebral artery, occipital artery** and **suboccipital** and **greater occipital nerves** (pp. 215–216). There are four suboccipital muscles that can be identified deep to the semispinalis capitis. The semispinalis should be reflected inferiorly by cutting its attachment from the occipital bone, being careful to protect the greater occipital nerve deep to this muscle. On a skeleton, identify parts of the **atlas** (**posterior tubercle, posterior arch** and **transverse process**) and **axis** (**spinous** and **transverse processes** and **odontoid**) and the occipital bone (nuchal lines). The **rectus capitis posterior major** muscle is the medial boundary of the triangle and extends between the spine of the axis and the inferior nuchal line. The **obliquus capitis superior** muscle forms the superior boundary of the triangle and attaches to the tip of the transverse process of the atlas and the occipital bone. The inferior boundary is formed by the **obliquus capitis inferior** muscle and extends between the spine of the axis and the transverse process of the atlas. The **rectus capitis posterior minor** is medial and deep to the rectus major and lies outside the triangle. After these four muscles have been identified, a number of relationships can be observed. All four muscles are innervated by the **suboccipital nerve** (dorsal ramus of the first cervical nerve), which can be found coursing between the **vertebral artery** and the **transverse arch of the atlas.** The **vertebral artery** can be identified on the superior surface of the posterior arch of the atlas after removal of the atlanto-occipital membrane, which forms the floor of the suboccipital triangle. The **greater occipital nerve** (dorsal ramus of the second cervical nerve) can now be seen emerging inferior to the inferior surface of the obliquus capitis inferior. This nerve provides sensory innervation to the scalp. The **occipital artery** can be identified lateral to the obliquus capitis superior.

8. VERTEBRAL COLUMN AND SPINAL CORD

Use your textbook, atlas and skeleton to review the general plan and characteristics of the vertebral column and to identify the specific regional characteristics of each vertebral region (pp. 175–179; 181–184). To study and examine the spinal cord, the vertebral canal will be opened from the seventh cervical vertebra to the lower lumbar vertebrae. First, remove the spinalis muscle on each side of the midline with the scalpel to expose each spinous process and lamina. With a chisel and mallet, perform a laminectomy by chiseling through each lamina about 1 cm on each side of the midline from C7 superiorly to the level of the lower lumbar vertebrae inferiorly. Place the chisel on each lamina at a 45-degree angle (Figure 2.5). Use the bone cutters and scalpel to trim away pieces of muscle and vertebrae to expose and widen the vertebral canal. After removing the lumbar laminae, identify the **ligamenta flava** (p. 182), which connect the laminae of adjacent vertebrae on their deep, inner surfaces. Next, carefully examine the surface of the dural sac by removing the fatty tissue from the epidural space. Note the extensive **internal vertebral venous plexus** in this epidural fat. Understand the importance of this venous plexus in the spread of cancer. The **dura mater** (p. 208) of the spinal cord is the outer meningeal layer and appears as a white, fibrous covering of the cord. The dura extends from the base of the skull to the level of the second sacral vertebra, where the filum terminale externum (coccygeal ligament) is formed. Review in your textbook the structure and extent of the filum below the second sacral vertebra. Open the dura with a midline vertical incision and carefully reflect it laterally, noting the extent of the dural sac. Deep to the dura, the **arachnoid membrane** can be seen as a thin, translucent structure collapsed over the spinal cord. Tease away the arachnoid membrane to observe the subarachnoid space, surface of the spinal cord and **pia mater,** which adheres tightly to and cannot be separated from the cord.

Identify the exits of several **dorsal** and **ventral roots** of the spinal nerves from the cord (pp. 208–209). Extensions of the dura, arachnoid and pia continue distally over the formation of each spinal nerve. Open several intervertebral spaces more laterally and care-

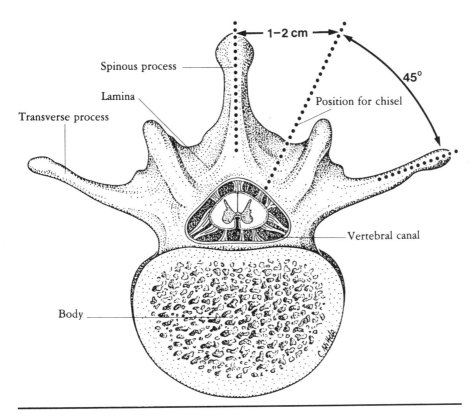

Figure 2.5

fully incise the dural sleeve over the dorsal and ventral roots. Observe the **spinal (dorsal root) ganglion** (p. 209), which forms an enlargement on the dorsal root at the intervertebral foramen. More laterally, the fusion of the roots forms the **spinal nerve,** which then divides into **dorsal** and **ventral rami** (pp. 208–209). Between the ventral and the dorsal roots in the dural sac and extending along each side of the cord, identify the **denticulate ligament,** which is a lateral extension of the pia mater (p. 209). Note that this ligament anchors to the dura at regular intervals between the exits of the ventral and dorsal roots from the dural sac. The parts of the sympathetic nervous system related to the spinal nerve complex will be dissected later with the thorax.

The spinal cord extends inferiorly to the level of the second or third lumbar vertebra. At this distal end, the cord tapers to form the **conus medullaris,** from which the pia mater continues as a thin strand (filum terminale internum) surrounded by the **cauda equina** (p. 208). The cauda equina is the massive collection of the ventral and dorsal roots of the spinal nerves within the inferior part of the dural sac (formed by the dura mater and arachnoid). Also, note the **cervi-**

cal and **lumbar enlargements** on the cord, which represent the origins of many nerves that supply the upper and lower limbs (pp. 208–209). Review the blood supply to the cord. In the lower thoracic region, cut transversely and remove a 5-cm segment of the cord to review the relationships described in the preceding text. Finally, note the **posterior longitudinal ligament** (p. 182) firmly anchored to the posterior surface of the vertebral bodies after removing this segment of the cord. Describe the functions and positions of the posterior and anterior longitudinal ligaments.

CHAPTER THREE
UPPER LIMB

9. REVIEW OF THE SUPERFICIAL MUSCLES OF THE BACK

The superficial muscles of the back functionally belong to the upper limb and were dissected with the back. These muscles (trapezius, latissimus dorsi, rhomboideus major and minor and levator scapulae) should be reviewed at this time (see Chapter 2; Section 6, Subsection A).

10. POSTERIOR SHOULDER AND DELTOID REGIONS

The muscles of the scapular region also functionally belong to the upper limb and are limb muscles that migrated to the back from the upper limb bud. With the cadaver in a prone position, extend the skin incisions (Figure 3.1) from the acromion at the posterior shoulder inferiorly to the posterior midpoint of the arm. Reflect the skin toward the ventral surface of the arm. Carefully remove the tela subcutanea, noting the cutaneous vessels and nerves in this fascia. Some of these nerves will be discussed and dissected later.

Clean the fascia from the lateral and posterior surfaces of the **deltoid** muscle to define these borders of the muscle (pp. 354–355). Watch for branches of the **superior lateral brachial cutaneous nerve** (pp. 373–374) at its posterior border. Cut the origin of the deltoid muscle from the spine and acromion of the scapula, but leave the muscle attached anteriorly to the clavicle. Also be sure that the insertions of the trapezius on the spine and acromion have been reflected. Elevate the posterior aspect of the deltoid and carefully remove the loose fascia deep to the posterior surface of the muscle with blunt dissection to identify the **axillary nerve** and **posterior circumflex humeral vessels** (Figure 3.2) within the fascia of the quadrangular space (pp. 373–374). Identify the **quadrangular and triangular spaces** and observe their boundaries and contents. Clean and define the **teres minor** and **major** muscles, which form the superior and interior borders, respectively, of the quadrangular and triangular spaces. Isolate the small branch of the axillary nerve that supplies the teres minor. Also, identify the **long head of the triceps,** which is closely related to the posterior border of the deltoid. Separate the long head from the surrounding scapular muscles, noting its relationships to the quadrangular and triangular spaces (p. 373).

The next step in your dissection (Figure 3.2) will be to identify the **supraspinatus** or **infraspinatus** muscles (pp. 355; 374). First, remove any remaining fascia deep to the trapezius and review the accessory nerve (subtrapezial plexus) and transverse cervical ves-

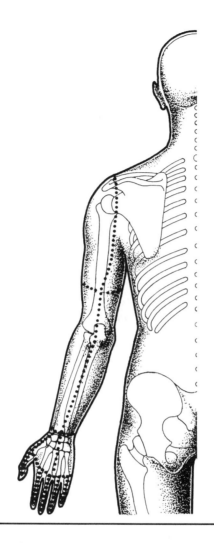

Figure 3.1

sels previously observed attached to the deep surface of the trapezius. Remove the fascia from the surface of the supraspinatus and identify the muscle deep to the trapezius in the supraspinatous fossa. At the superior border of the scapula, palpate this margin of the bone between the attachment of the levator scapulae medially and the **scapular notch** laterally. With the scalpel, carefully reflect the origin of the supraspinatus from the medial end of the supraspinatous fossa. Then, with blunt dissection, continue to pull the muscle laterally. Saw through the acromial process of the scapula (dotted lines in Figure 3.2) at its junction with the spine to make it easier to pull the supraspinatus laterally toward its insertion on the greater tubercle of the humerus. With force, push the acromion anteriorly. As the

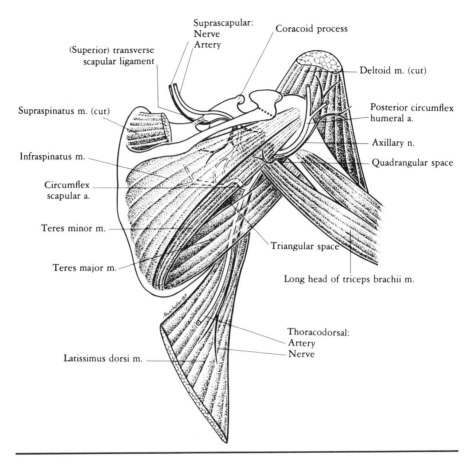

Figure 3.2

supraspinatus is pulled laterally, identify the **suprascapular nerve and vessels** on its deep surface at the scapular notch. The suprascapular nerve innervates the supraspinatus and infraspinatus. Identify the (**superior**) **transverse scapular ligament** as it crosses the scapular notch (p. 343) and note that the nerve usually passes deep to and the vessels superior to this ligament (Figure 3.2). Follow the suprascapular nerve and vessels proximally for a short distance into the fascia deep to the trapezius. Note the origin of the **inferior belly of the omohyoid** muscle adjacent to the scapular notch. Follow the suprascapular nerve and vessels distally deep to the supraspinatus. The nerve and vessels pass around the notch of the scapular neck to reach the deep surface of the infraspinatus in the infraspinatous fossa, as described in the text that follows. In the infraspinatous fossa, clear the fascia from the superficial surface of the infraspinatus and note the direction of its fibers and the dense nature of the fascia.

To follow the suprascapular nerve and vessels into the infraspinatous fossa, cut the fibers of origin of the infraspinatus from their medial attachments on the dorsum of the scapula and reflect the muscle laterally toward its insertion on the greater tubercle of the humerus. Note the close relationship between the infraspinatus and the teres minor muscles and take care to separate the teres minor from the infraspinatus. Do not reflect the teres minor. Review the branch of the axillary nerve to the teres minor, noting its insertion also on the greater tubercle (pp. 345; 374). Inferior to the teres minor, again identify the teres major and review the relationships of the long head of the triceps to the teres major and minor muscles (p. 374). Identify the **circumflex scapular vessels** as they pass around the lateral border of the scapula in the triangular space and the **axillary nerve** and **posterior circumflex humeral vessels** in the quadrangular space (p. 374). Review the relationships of the supraspinatus, infraspinatus and teres minor to the shoulder joint. Review the posterior course and insertion of the **serratus anterior.**

11. CUTANEOUS VESSELS AND NERVES

The skin on the anterior chest wall and upper limb will now be reflected. Turn the body to the supine position and make the skin reflections on the ventral side of the chest and arm, as indicated in Figure 3.3. Cut the skin around the nipples. Leave the superficial fascia intact at this time. Reflect all the skin from the lateral chest wall and armpit. As the skin is reflected at the clavicle, note the inferior fibers of the **platysma** muscle in the superficial fascia. The platysma is a muscle of facial expression that will be studied with the head and neck. On the arm, reflect the cut edges of the skin incisions posteriorly around the limb. Retain the skin flaps to lay over the dissections after each laboratory period.

With the skin reflected, remove the superficial fascia on the anterior chest wall and upper arm. Review the cutaneous nerves and veins described in the text that follows. In the female, the mammary gland shoud be removed en masse with the superficial fascia. It is not necessary to dissect into the gland because it is composed primarily of fatty and connective tissue in most older cadavers. Review the lymphatic drainage of the mammary gland, noting its importance in the spread of cancer. Note that the entire gland is superficial to the deep fascia that covers the pectoralis major muscle. Identify examples of lateral and anterior cutaneous nerves and vessels during removal of the tela subcutanea; most of these nerves and vessels are branches of the intercostal nerves and arteries (pp. 377–378).

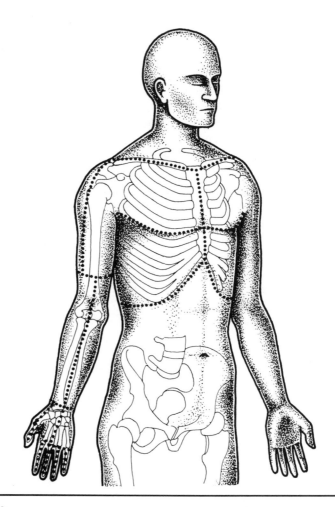

Figure 3.3

The cutaneous veins are numerous in the superficial fascia of the limb. While removing the superficial fascia, identify the **cephalic vein** at the lateral aspect of the arm and the **basilic vein** at the inferior, medial side of the arm (pp. 377; 385; 390). The cephalic and basilic veins begin at the dorsal venous arch on the posterior aspect of the hand. The cephalic vein is the lateral continuation of the arch and the basilic vein is formed as the medial continuation. The cephalic vein ascends the lateral surface of the forearm, passes anterior to the elbow and ascends the lateral side of the arm to the anterior shoulder region between the deltoid and the pectoralis major muscles (Figure 3.4). At the clavicle, the cephalic vein penetrates the costocoracoid membrane to terminate deeply in the axillary vein in the

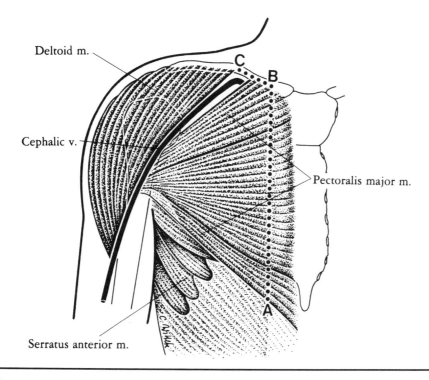

Figure 3.4

axilla. The basilic vein ascends the medial side of the forearm and passes ventral to the medial epicondyle at the elbow, piercing the brachial fascia at the lower third of the arm. It then passes as a deep vein through the neurovascular compartment of the arm to reach the axilla, where it joins the brachial veins to form the axillary vein. The median cubital vein connects the cephalic and basilic veins superficially in the cubital fossa at the elbow.

The cutaneous nerves of the upper limb are primarily either direct or indirect branches of the brachial plexus. These nerves are numerous throughout the superficial fascia of the limb, making identification of all of them unnecessary in the laboratory. As the superficial fascia is removed during subsequent dissections of the upper limb, some of the following major cutaneous nerves can be identified:

1. Posterior brachial
2. Medial brachial
3. Intercostobrachial
4. Lateral brachial
5. Medial antebrachial
6. Lateral antebrachial
7. Posterior antebrachial
8. Superficial radial

12. PECTORAL REGION

Dissection of the pectoral region will complete the study of the muscles and associated structures of the upper limb that migrated to the trunk. These muscles are located anteriorly on the chest wall with origins from the ribs, sternum or clavicle and insertions on the humerus and scapula.

Remove the deep fascia from the surface of the **pectoralis major** muscle to identify its borders (pp. 356; 377). Identify the clavicular and sternal origins of the pectoralis major (Figure 3.4) and note that its fibers narrow laterally, forming the anterior axillary fold as it passes to insert on the **crest of the greater tubercle of the humerus.** Before reflecting this muscle, identify the anterior border of the **deltoid** and separate it from the pectoralis major. Also, preserve and follow the superior continuation of the cephalic vein between these two muscles, noting its deeper course to the pectoralis major to terminate in the axillary vein, as described in the text that follows. Reflect the pectoralis major laterally by cutting its fibers of origin from the sternum, ribs and clavicle. Begin this reflection at the inferior sternal fibers (Figure 3.4, A-B) and continue superiorly to the clavicle. Be careful when cutting the clavicular fibers (Figure 3.4, B-C) because immediately deep to these fibers, the branches of the **thoraco-acromial arterial trunk, lateral pectoral nerve** and deep terminal course of the **cephalic vein** can be identified (pp. 378; 381). Note that all these structures are medial to the pectoralis minor and penetrate the costocoracoid membrane of the clavipectoral fascia (described later). The thoracoacromial trunk (Figure 3.5) is a branch of the second part of the axillary artery. Identify this arterial trunk in the fascia deep to the clavicular fibers of the pectoralis major and observe its division into four branches: (1) **pectoral branches** to the deep surface of the pectoralis major. The pectoral branches can be traced deeply to locate the thoracoacromial trunk. (2) The **acromial branch** passes deep to the anterior border of the deltoid to supply muscles at the acromion. (3) The **deltoid branch** (often a branch of the acromial artery) passes between the deltoid and the pectoralis major and accompanies the cephalic vein and supplies these two muscles. (4) The **clavicular branch** supplies the **subclavius** muscle (identified on the deep surface of the clavicle) and sternoclavicular joint (p. 381). Each of these branches is extremely small and must be carefully dissected from the fascia. Venous branches accompany each of these arterial branches and are tributaries to the axillary vein. Traveling with the pectoral vessels medial to the pectoralis minor is the lateral pectoral nerve, which innervates the pectoralis major (p. 378). This nerve can later be followed to the lateral cord of the

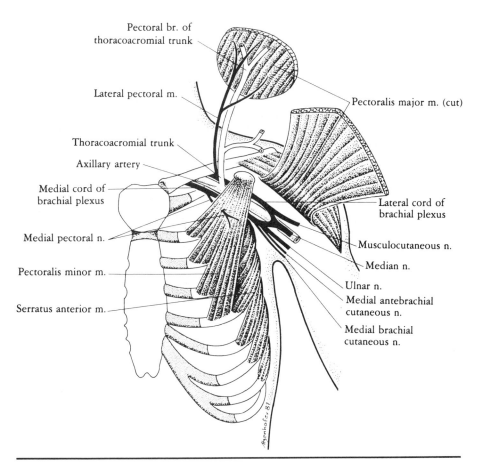

Pectoral br. of
thoracoacromial trunk

Lateral pectoral m.

Pectoralis major m. (cut)

Thoracoacromial trunk

Axillary artery

Medial cord of
brachial plexus

Lateral cord of
brachial plexus

Medial pectoral n.

Musculocutaneous n.

Median n.

Pectoralis minor m.

Ulnar n.

Medial antebrachial
cutaneous n.

Serratus anterior m.

Medial brachial
cutaneous n.

Figure 3.5

brachial plexus. To facilitate complete reflection of the pectoralis
major to its insertion on the humerus, cut out a cube of the muscle
with the pectoral vessels and lateral pectoral nerve attached (Figure
3.5). Follow the terminal course of the cephalic vein to the axillary
vein.

After reflecting the pectoralis major laterally, note the position
and clean the superior (medial) and inferior (lateral) borders of the
pectoralis minor muscle (Figure 3.5), which is immediately deep to
the pectoralis major (pp. 357; 378). The pectoralis minor is enclosed
in its deep fascia, the **clavipectoral fascia.** This continuous layer of
deep fascia covers the pectoralis minor and extends from the clavicle
superiorly to the axillary fascia inferiorly. The axillary fascia forms
the base of the axilla and is continuous with the fascia of the latissimus dorsi posteriorly and the fascia of the pectoralis major ante-

riorly. The part of the clavipectoral fascia between the clavicle and the pectoralis minor is the **costocoracoid membrane.** This structure is penetrated by the cephalic vein, lateral pectoral nerve and thoracoacromial trunk, as described previously. Inferior to the pectoralis minor and extending between this muscle and the axillary fascia, the clavipectoral fascia is known as the **suspensory ligament of the axilla,** which helps support the axillary fascia. Follow the fibers of the pectoralis minor superiorly to their insertions on the **coracoid process of the scapula.** The pectoralis minor is triangular and should be reflected at its midpoint between the ribs and the coracoid process (p. 381). Note that the **medial pectoral nerve** usually pierces the pectoralis minor (Figure 3.5) or passes around its lateral (inferior) border to innervate the pectoralis minor. The medial pectoral nerve then continues superficially to innervate the pectoralis major.

13. AXILLA

The axilla is the pyramidal area of transition between the trunk and the upper limb. Through the axilla pass all the major nerves, blood vessels and lymphatic vessels to and from the upper limb. Like any anatomic area, the axilla should first be described by its bony and muscular boundaries; then, its contents should be identified and their relationships described. Follow this study pattern as you proceed to other anatomic areas.

Use your textbook and dissection to review the structures that form the boundaries of the axilla. The pectoralis major and minor muscles and clavipectoral fascia form the anterior wall. With these two muscles already reflected, the contents of the axilla are exposed and ready for dissection. The **serratus anterior** muscle (Figure 3.5) and upper ribs form the medial boundary (p. 381). The subscapularis, latissimus dorsi and teres major comprise the posterior boundary (pp. 358–359). Laterally, the converging fibers of the latissimus dorsi and pectoralis major form a narrow lateral boundary, the intertubercular groove of the humerus. The base of the axilla is formed by the axillary fascia and the apex is formed by the converging borders of the clavicle, scapula and first rib (p. 344).

To display the major contents of the axilla and to observe their relationships, carefully remove the loose fascia, lymph nodes and fat that surrounds the **axillary vein, axillary artery** and **brachial plexus.** The vessels and nerves are enclosed in a fascial covering, the **axillary sheath,** which must be teased away.

A. Axillary Vein

The **axillary vein** and its tributaries are large and numerous (p. 370). This vein is superficial and overlies the deeper axillary artery, often obscuring many of its branches. The vein begins at the lateral border of the teres major by the junction of the two brachial veins and basilic vein. Superiorly, the axillary vein becomes the subclavian vein as it crosses the lateral border of the first rib. Its many tributaries correspond to branches of the axillary artery and should be identified as the artery is dissected. If the vein is unusually large, it may be necessary to remove a large segment of it and its tributaries between the base and the apex of the axilla to provide a better view of the axillary artery and brachial plexus.

B. Axillary Artery

The **axillary artery** is the main source of blood to the upper limb and shoulder (pp. 368–369; 381–384). As a segment of the main arterial flow from the heart, it extends from the lateral border of the first rib to the inferior border of the teres major. The artery is conveniently divided into three parts by the overlying position of the pectoralis minor (Figure 3.6). The first part is proximal (medial) to the pectoralis minor, the second part deep to this muscle and the third part distal (lateral) to it. Many variations of the typical vascular patterns in the axilla may occur during development. Therefore, it is imperative that the *entire* distributions of the arteries be displayed by blunt dissection to allow observation of the final distribution of the artery from its point of origin to allow correct identification. Careful dissection should permit identification of a number of branches from the three parts of the axillary artery (Figure 3.6).

First Part. There is usually one artery from the first part, the **superior (supreme) thoracic artery** (pp. 381–382). This small artery supplies the first and second intercostal spaces. It is often absent or easily broken.

Second Part. There are two branches from the axillary artery deep to the pectoralis minor (pp. 381–382). The first branch is the **thoracoacromial trunk,** which passes around the medial border of the pectoralis minor. Its branches have previously been dissected and should be reviewed.

The second branch is the **lateral thoracic artery.** This artery is a more variable branch, originating from many possible sources. It is usually identified at the lateral border of the pectoralis minor as it

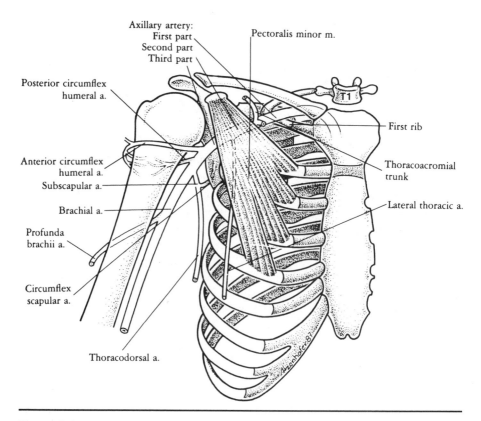

Figure 3.6

courses on the superficial surface of the **serratus anterior** to supply structures of the chest wall. The lateral thoracic artery is accompanied in part by the **long thoracic nerve.**

Third Part. The third part of the artery is distal (lateral) to the pectoralis minor and courses parallel to the coracobrachialis muscle of the arm. To provide a better field of dissection, separate the fascia and tissues of the upper arm for a short distance along the courses of the axillary and brachial arteries, thereby establishing better continuity between the axilla and the arm. From the third part, three arteries arise (pp. 381–384).

The first branch is the **subscapular artery.** This large trunk can be found at the lateral border of the **subscapularis** muscle. The subscapular artery quickly divides into two branches. First is the **circumflex scapular artery,** which passes posteriorly around the lateral border of the scapula through the triangular space. The circumflex artery supplies the dorsal scapular muscles and anastomotic scapular branches to the suprascapular and dorsal scapular arteries. The second branch of the subscapular artery is the **thoracodorsal**

artery, the main blood supply to the latissimus dorsi. Dissect the distributions of these vessels through the fascia and observe their relationships.

The second branch from the third part of the axillary artery is the **posterior circumflex humeral artery.** This artery arises from the side of the axillary artery opposite the origin of the subscapular artery. It passes dorsally around the surgical neck of the humerus with the **axillary nerve** through the quadrangular space.

The third branch is the smaller **anterior circumflex humeral artery,** which often arises with the posterior circumflex humeral artery. It completes the formation of the arterial circle around the surgical neck of the humerus deep to the brachial muscles. It is smaller than the posterior circumflex humeral artery.

C. Brachial Plexus

One of the most rewarding dissections of the axilla is that of the **brachial plexus** and its branches. The brachial plexus is the complex organization of nerves that supply motor and sensory innervation to the upper limb, most of the shoulder and most of the superficial back. The basic structural pattern of the brachial plexus is described in terms of rami (ventral rami of the fifth cervical to the first thoracic spinal nerve), trunks, divisions, cords and terminal branches. The plexus originates from the ventral rami of the fifth cervical to the first thoracic spinal nerve. The rami join to form the superior, middle and inferior trunks. The rami and trunks are located superior to the clavicle (p. 383) (supraclavicular part) in the posterior triangle of the neck. The divisions (anterior and posterior) are deep to the clavicle. Therefore, these superior parts of the plexus will be dissected later with the neck. The cords and terminal branches are found in the axilla inferior to the clavicle (infraclavicular part) (pp. 382–384) and will now be dissected. The anterior and posterior divisions represent the separation of nerve fibers that innervate the ventral and dorsal muscle masses of the limb. The divisions fuse to form the posterior, lateral and medial cords in the axilla, each named according to its relationship to the second part of the axillary artery deep to the pectoralis minor muscle. The posterior cord is formed by the junction of the three posterior divisions derived from the three trunks. The posterior cord innervates muscles of the dorsal (extensor) surface of the limb. The anterior divisions of the superior and middle trunks form the lateral cord, whereas the anterior division of the inferior trunk forms the medial cord. The latter two cords and their branches innervate the ventral (flexor) surface of the limb.

Identify the **lateral, medial** and **posterior cords** (Figure 3.5) that surround the second part of the axillary artery (pp. 383–384). Elevate the axillary artery to observe the posterior cord deep to the artery. The lateral and medial cords can be seen on their respective sides of the artery. From the lateral and medial cords, locate the **lateral** and **medial pectoral nerves** and follow their innervations to their respective muscles. The lateral pectoral nerve passes medial to the pectoralis minor to reach the pectoralis major. The medial pectoral nerve passes between the axillary artery and the axillary vein to reach the pectoralis minor and major. From the medial cord, the origins of three additional branches can be dissected. Two cutaneous branches are the **medial brachial cutaneous nerve** to the medial side of the arm and the **medial antebrachial cutaneous nerve** to the medial aspect of the forearm. The latter nerve follows the basilic vein distally. The third and largest branch of the medial cord is the **ulnar nerve,** which innervates muscles and skin of the forearm and hand. Follow the ulnar nerve from the axilla to the point where it enters the neurovascular compartment of the arm, where it will be dissected.

Identify the **median nerve** and note that it is formed by contributions from the medial and lateral cords anterolateral to the axillary artery (Figure 3.5). The median nerve supplies part of the flexor surface of the forearm and hand. Follow the median nerve into the arm. From the lateral cord, identify the **musculocutaneous nerve** distal to the contribution of the lateral cord to the median nerve. The musculocutaneous nerve innervates the ventral (flexor) muscles of the arm and lateral skin of the forearm. Follow this nerve to its point of entry into the coracobrachialis muscle.

From the posterior cord, nerves are formed that supply the dorsal muscle mass and skin of the shoulder and limb. From the middle part of the posterior cord on the anterior surface of the subscapularis, identify the **upper, middle (thoracodorsal)** and **lower subscapular nerves** from the fascia that covers the subscapularis muscle. All these nerves supply the subscapularis and can be variable in number. The thoracodorsal nerve also supplies the latissimus dorsi, whereas the lower subscapular nerve supplies the teres major. The middle and lower nerves often have a common origin. Clean the surface of the **subscapularis** muscle to identify these nerves and muscle. Follow completely the thoracodorsal nerve as it travels with the thoracodorsal vessels to their terminations on the latissimus dorsi. Distal to these three subscapular nerves, the **axillary nerve** branches from the posterior cord (pp. 382–383). The lower supscapular nerve often arises from the axillary nerve. Free the axillary nerve from the fascia of the subscapularis and follow its course distally until it passes around the lower border of this muscle through the quadrangular space with the posterior circumflex humeral artery. The axillary nerve innervates the deltoid and teres minor and provides cutaneous nerves to the posterior shoulder. Distal to the axillary nerve, the posterior cord continues as the **radial nerve,** the sole innervation of the dorsal muscle mass and skin of the upper arm and forearm (p. 384). Follow the radial nerve into the arm.

Note that the origins of the musculocutaneous, median and ulnar nerves form the characteristic M configuration on the ventral surface of the axillary artery (p. 384). Also, observe that the musculocutaneous nerve is lateral to the axillary artery, whereas the ulnar nerve and medial brachial and antebrachial cutaneous nerves are medial to the artery in the axilla. The median nerve is initially anterolateral to the axillary artery and the radial nerve spirals posterior to the artery to enter the radial groove of the humerus. These terminal nerves will be followed in more detail during the dissection of the remainder of the limb.

D. Axillary Lymph Nodes

The 20 to 30 nodes (pp. 379–380) within the axilla are arranged in five groups: lateral, subscapular, pectoral, central and apical. Although these nodes may not be readily identified in the dissection, it is important to understand that they receive afferent lymphatic vessels from the upper limb, shoulder and, most importantly, the mammary gland (especially to the pectoral group). Collectively, these nodes form the subclavian lymphatic trunk, which drains into the junction of the right subclavian and internal jugular veins or the left brachiocephalic vein.

14. ARM

On a dried bone, identify the **greater** and **lesser tubercles** and **intertubercular groove** on the proximal end of the humerus (p. 345). On the shaft, note the **radial groove.** Observe the **medial** and **lateral epicondyles** at the distal ends of the humerus. The medial epicondyle is most prominent. Note the **supracondylar ridges** above each epicondyle. On the ventral surface of the distal humerus between the condyles are the rounded **capitulum** and spool-shaped **trochlea.** The **olecranon** process of the ulna can be seen posteriorly.

The arm is the segment of the upper limb between the shoulder and the elbow. Most of the structures dissected in the axilla pass into the upper arm and continue into the forearm and hand. The deep structures of the arm are located primarily in two fascial compartments: the anterior (flexor) and posterior (extensor) compartments, formed by the brachial fascia and intermuscular septa. Remove the ventral skin and superficial fascia from the arm to the wrist (Figure 3.2). Review the courses of the median cubital, cephalic and basilic veins and observe several of the major cutaneous nerves that course in the superficial fascia. Deep to the superficial fascia are the circular fibers of the deep brachial fascia, which supports and maintains the structures of the arm. Deep extensions of the brachial fascia to the lateral and medial aspects of the humerus form the **lateral** and **medial intermuscular septa** that define the anterior and posterior compartments (pp. 355–359). These septa are most prominent in the lower half of the arm and can be identified distal to the insertions of the deltoid and coracobrachialis.

A. Anterior (Flexor) Compartment

With scissors, carefully open the anterior compartment with a ventral, longitudinal cut through the brachial fascia. There are three muscles in the anterior compartment: the biceps brachii, brachialis and coracobrachialis. First, identify the **biceps brachii** (Figure 3.7), noting its **long** and **short heads** of origin (pp. 358–359; 382). This muscle spans the ventral surface of the arm and its tendon of insertion will be dissected later as it passes through the cubital fossa to insert on the tuberosity of the radius (pp. 347; 386). At the cubital fossa, note the superficial, medially directed, aponeurotic extension (**bicipital aponeurosis**) from the medial side of the biceps tendon. This aponeurosis provides a protective covering for the median nerve and brachial artery at the cubital fossa (p. 385). Clean the superficial structures of the cubital fossa and cut the bicipital aponeurosis to note the relationships of the brachial artery and median nerve to the biceps tendon in the cubital fossa.

In the middle of the arm, free and retract the biceps from the loose fascia and identify the **brachialis** (Figure 3.8) deep to the inferior half of the biceps (pp. 358–359). The brachialis covers the ventral surface of the inferior humerus. Its tendon of insertion will be followed later to the tuberosity of the ulna in the floor of the cubital fossa (p. 387). The third muscle in the anterior compartment is the **coracobrachialis** (Figure 3.7), which is medial and parallel to the short head of the biceps (p. 359). This muscle originates from the coracoid process of the scapula and courses parallel to the axillary and brachial arteries. The coracobrachialis inserts on the medial aspect of the midpoint of the humerus. All three anterior muscles are innervated by the **musculocutaneous nerve.** This nerve arises from the lateral cord of the brachial plexus and pierces the coracobrachialis.

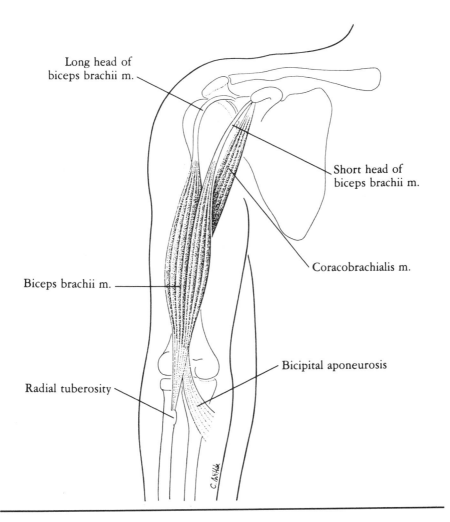

Long head of
biceps brachii m.

Short head of
biceps brachii m.

Coracobrachialis m.

Biceps brachii m.

Bicipital aponeurosis

Radial tuberosity

Figure 3.7

Follow the musculocutaneous nerve through this muscle to travel
in the muscular plane between the brachialis and the biceps (Figure 3.8). Distally on the lateral aspect of the biceps, note that the
musculocutaneous nerve becomes the **lateral antebrachial cutaneous nerve,** which enters the superficial fascia of the lateral forearm
after emerging between the biceps and the brachialis (p. 231).

The **brachial artery** (p. 384) is the continuation of the axillary
artery distal to the inferior border of the teres major. It courses
through the flexor compartment of the arm within the neurovascular
compartment (formed by the medial intermuscular septum). The brachial artery is superficial during most of its course, being at first just
deep to the skin and fascia. The proximal segment of this artery is

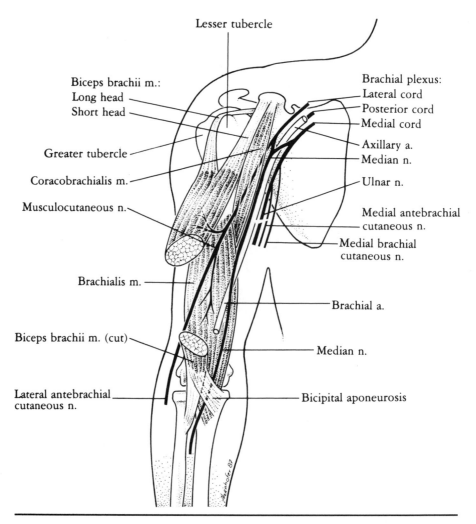

Lesser tubercle

Biceps brachii m.:
Long head
Short head

Greater tubercle

Coracobrachialis m.

Musculocutaneous n.

Brachialis m.

Biceps brachii m. (cut)

Lateral antebrachial
cutaneous n.

Brachial plexus:
Lateral cord
Posterior cord
Medial cord

Axillary a.
Median n.

Ulnar n.

Medial antebrachial
cutaneous n.
Medial brachial
cutaneous n.

Brachial a.

Median n.

Bicipital aponeurosis

Figure 3.8

just medial to the biceps and coracobrachialis. In the upper half of the arm, the **ulnar nerve** is medial to the artery and the **median nerve** is initially lateral to it. During its course, the median nerve passes from the lateral to the medial side of the artery. After crossing the elbow joint, the brachial artery divides into the radial and ulnar arteries within the cubital fossa, which will be dissected later. Note that there are usually two **brachial veins** (venae comitantes) that accompany the brachial artery. They terminate superiorly by joining with the basilic vein to form the axillary vein at the inferior border of the teres major.

Many unnamed muscle arteries and several nutrient arteries to the humerus arise from the brachial artery. Three named branches should be identified:

1. The **profunda brachii artery,** or **deep femoral artery** (p. 384), is the highest branch that arises just inferior to the teres major. This artery can be extremely small and courses around the medial side of the humerus to enter the radial groove and posterior compartment on the dorsum of the arm with the radial nerve.

2. From the middle segment of the brachial artery, the **superior ulnar collateral artery** (p. 369) courses medially and penetrates the medial intermuscular septum to pass from the anterior to the posterior compartment. It descends with the ulnar nerve in the posterior compartment of the arm and passes posterior to the medial epicondyle of the humerus, where it participates in the collateral circulation around the elbow joint.

3. From the distal end of the brachial artery immediately superior to the condyles, the **inferior ulnar collateral artery** (pp. 386–387) passes anterior to the medial condyle to anastomose with vessels of the forearm at the elbow.

Follow the **median nerve** from its origins from the lateral and medial cords of the brachial plexus (Figure 3.8). It lies first on the lateral side of the brachial artery as it enters the arm from the axilla. Descending through the anterior compartment, it courses ventrally across the brachial artery in the middle of the arm and is found on the medial side of the artery in the cubital fossa, where it enters the forearm (p. 384). The biceps tendon (p. 386) is on the lateral side of the artery in the cubital fossa.

The **ulnar nerve** (p. 384), a branch of the medial cord in the axilla, enters the arm on the medial side of the brachial artery. Midway through the arm, it pierces the medial intermuscular septum to enter the posterior compartment of the arm. Traveling with the superior ulnar collateral artery, the ulnar nerve enters the forearm posterior to the medial epicondyle. Neither the median nor the ulnar nerve provides any branches in the arm.

B. Posterior (Extensor) Compartment

Complete the removal of the skin and fascia on the dorsum of the upper limb as far as the wrist, noting cutaneous nerves (Figure 3.1) within the extensor compartment. Identify the **triceps brachii,** which is the large posterior muscle mass of the arm (pp. 354–355; 358–359). There are three heads of origin to be identified (Figures 3.9 and 3.10). The **long head** arises from the **infraglenoid tubercle** of

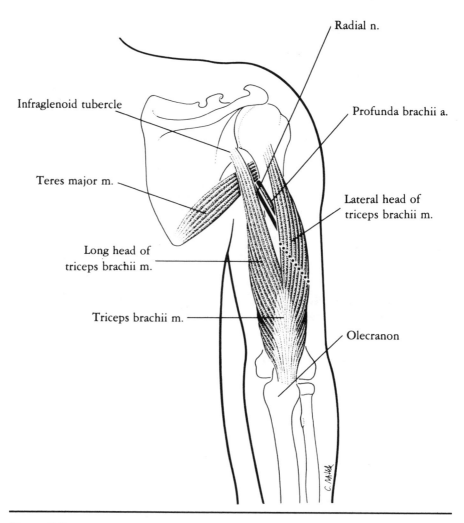

Figure 3.9

the scapula. Review the relationships, of the long head to the teres major and minor (p. 373). The **medial** and **lateral heads** arise from the humerus and are separated by the radial groove. The **lateral head** lies superior and lateral to the groove, whereas the **medial head** lies inferior and medial to it. The medial head is visualized better after an oblique cut is made through the lateral head. Make this cut (dotted line in Figure 3.9) from the superior, medial border of the lateral head to its inferior, lateral border to display the radial groove (p. 374). This cut parallels the course of the radial nerve and profunda brachii vessels deep to the lateral head. Avoid damaging these vessels and nerve deep to the muscle. Cut only the lateral head

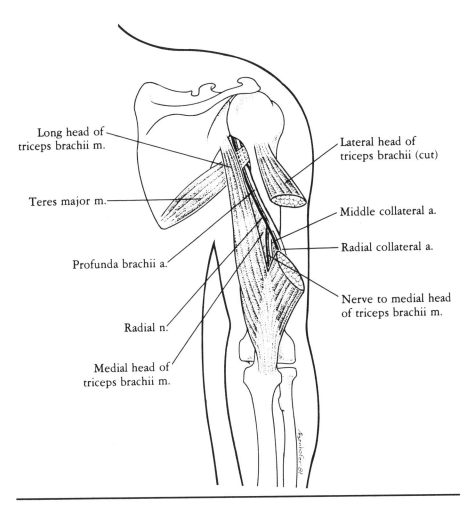

Long head of
triceps brachii m.

Teres major m.

Profunda brachii a.

Radial n.

Medial head of
triceps brachii m.

Lateral head of
triceps brachii (cut)

Middle collateral a.

Radial collateral a.

Nerve to medial head
of triceps brachii m.

Figure 3.10

of the muscle and, with it reflected, dissect the course of the **radial nerve** through the posterior compartment of the arm (Figure 3.10). The radial nerve takes a spiral course in the radial groove on the posterior surface of the humerus, where it gives branches to the three heads of the triceps brachii (p. 374). The radial nerve also provides posterior brachial and posterior antebrachial cutaneous branches to the arm and forearm, respectively, in the groove. In the lower third of the arm, the radial nerve pierces the lateral intermuscular septum and is just ventral to the lateral epicondyle of the humerus between the **brachialis** and the **brachioradialis** muscles. Separate these two muscles to observe the radial nerve at this location distal to the radial groove. Before entering the cubital fossa, the radial nerve innervates two extensor muscles of the forearm, the **brachioradialis** and **extensor carpi radialis longus.**

Joining the radial nerve proximal to the radial groove is the **profunda brachii artery** (Figure 3.10), which is a branch of the brachial artery (pp. 374; 384). Before entering the groove, this artery gives rise to an ascending branch that anastomoses with the anterior and posterior circumflex humeral vessels (p. 374), forming an important collateral channel between the axillary and the brachial arteries. In the radial groove, the profunda brachii (deep radial) artery divides to form the radial collateral and middle collateral (Figure 3.10) arteries (p. 374). These arteries are small and often difficult to dissect. The radial collateral artery follows the course of the radial nerve anterior to the lateral epicondyle into the forearm. The middle collateral artery penetrates the medial head of the triceps and passes posterior to the lateral epicondyle. Both arteries complement the collateral circulation at the elbow.

15. CUBITAL FOSSA

The cubital fossa (Figure 3.11) is the triangular region on the ventral surface of the elbow that contains vessels and nerves that pass between the arm and the forearm. Only the ulnar nerve enters the forearm by passing dorsal to the medial epicondyle, outside the cubital fossa. Superficially, the fossa is an easy site for withdrawing blood because the **median cubital vein** (p. 385) is in the superficial fascia. Review the superficial structures of the fossa. An imaginary line drawn through the two epicondyles forms the superior boundary of the fossa. The roof is formed by skin, superficial fascia and the bicipital aponeurosis (p. 385), which have been removed. The floor is formed by the insertion of the **brachialis** and a part of the **supinator** muscle (pp. 386–387; 391). The **brachioradialis** of the forearm is the lateral boundary, whereas the **pronator teres** muscle of the forearm forms the medial boundary. Identify the **radial nerve** on the lateral side of the fossa deep to the brachioradialis and note its division in the cubital fossa into **superficial** and **deep radial nerves.** The superficial nerve passes deep to the brachioradialis and the deep nerve penetrates the supinator. In the central part of the cubital fossa, identify the lateral to medial relationship of the **biceps brachii tendon, brachial artery** and **median nerve.** The brachial artery divides into the **radial** and **ulnar arteries** at the apex of the fossa. All these arteries will be followed into the forearm.

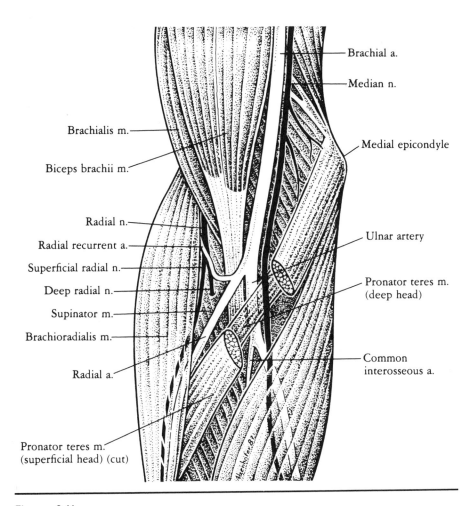

Brachial a.

Median n.

Brachialis m.

Biceps brachii m.

Medial epicondyle

Radial n.

Radial recurrent a.

Superficial radial n.

Deep radial n.

Supinator m.

Brachioradialis m.

Ulnar artery

Pronator teres m.
(deep head)

Common
interosseous a.

Radial a.

Pronator teres m.
(superficial head) (cut)

Figure 3.11

16. FOREARM

The forearm is the segment of the upper limb between the elbow and the wrist. It contains two bones, the **radius** and **ulna,** connected by an interosseous membrane. Use the skeleton and your atlas and textbook to study the main features of the radius and ulna (pp. 346–349). The forearm is divided into ventral (flexor and pronator) and dorsal (extensor and supinator) compartments similar to those of the arm. Most of the muscle masses of the two compartments form tendons that cross the wrist to enter the hand. Therefore, the dorsum of the hand will be dissected with the extensor compartment of the forearm and the palm of the hand will be dissected with the flexor compartment. Study and identify the bones of the wrist and hand (pp. 348–349), noting their general relationships and articulations with each other.

17. EXTENSOR COMPARTMENT OF THE FOREARM AND DORSUM OF THE HAND

Complete the removal of skin and superficial fascia from the dorsum of the hand and fingers distally to the nail beds (Figure 3.1). Identify, but do not cut at this time, the **extensor retinaculum,** a transverse thickening of deep fascia at the wrist (p. 366). In the superficial fascia on the dorsum of the hand, carefully observe the dorsal venous plexus that forms the **basilic** and **cephalic veins** (p. 390). It will be necessary to transect the middle of the venous plexus and retract it to the sides of the hand. Also, note the cutaneous nerves on the dorsum of the hand, which are derived from the radial (**superficial radial nerve**) and ulnar nerves (**dorsal cutaneous branch**) (p. 390). Preserve these two nerves during removal of the tela subcutanea at each side of the wrist.

The muscles of the posterior extensor compartment are arranged into superficial and deep groups. Most of the superficial muscles originate from the lateral epicondyle and supracondylar ridge of the humerus. The deep muscle group arises from the dorsal surfaces of the radius and ulna and the interosseous membrane. It is not necessary to bisect any of the extensor muscles to observe relationships, nerves or vascular structures.

A. Superficial Muscle Group

First, identify the muscles that comprise the superficial group (pp. 366; 391). From the lateral to the medial side of the forearm, they are the **brachioradialis, extensor carpi radialis longus, extensor carpi radialis brevis, extensor digitorum, extensor digiti minimi** and **extensor carpi ulnaris.** The first two muscles are innervated by the radial nerve. The extensor carpi radialis brevis is innervated by the deep radial nerve proximal to the supinator muscle. The others are innervated by the deep radial nerve (posterior interosseous nerve) (p. 391) distal to the supinator. Except for the brachioradialis, the tendons of these muscles cross the carpal bones at the wrist to reach the dorsal surface of the hand. Study the details of their insertions and actions on the hand and wrist in your textbook. The carpi muscles insert on the metacarpal bones and the digiti muscles form an **extensor expansion** on the dorsal surfaces of the second to fifth digits (p. 366). Following the description in your textbook, understand the structure of the extensor expansion. The detailed dissection of the expansion will be completed later in the hand.

B. Deep Muscle Group

In the deep compartment of the forearm, there are five muscles: the **supinator, abductor pollicis longus, extensor pollicis longus, extensor pollicis brevis** and **extensor indicis** (p. 367). The supinator does not enter the hand. Observe this muscle in the proximal forearm in the floor of the cubital fossa (p. 363). Define its origin and note that its fibers wrap around the radius laterally to insert on the anterior surface of the shaft of this bone inferior to the insertion of the pronator teres. Cut and separate the superior, common fibers of the extensor carpi radialis brevis and extensor digitorum for 4 to 5 cm superiorly toward the epicondyle to better identify the supinator. Note the course of its fibers and describe its functions. Observe that the **deep radial nerve** innervates and penetrates the supinator in the cubital fossa to reach the posterior compartment of the forearm, where it is called the posterior interosseous nerve (or deep radial nerve). Identify the remaining four muscles, which enter the hand and insert on the thumb or index finger. These four muscles are innervated by the deep radial nerve distal to the supinator. The three pollicis muscles cross superficial to the extensor carpi radialis longus and brevis and the brachioradialis to reach the lateral side of the thumb (p. 367). The extensor indicis is a separate tendon to the index finger that joins the extensor expansion of the second digit.

The tendons (except the brachioradialis) from both groups of muscles are bound to the carpal bones by the **extensor retinaculum** as they cross the wrist. Identify the retinaculum, which extends from the radius laterally to the styloid process of the ulna and the pisiform and triquetral bones medially (pp. 366–367). Deep to the retinaculum, the extensor tendons pass through six compartments. Cut and open the retinaculum to expose the courses of the tendons onto the hand and completely clean the tendons from the dorsal fascia distal to the retinaculum onto the digits. As the tendons pass deep to the retinaculum, they are covered by **synovial sheaths** (p. 366), which provide frictionless movement of the tendons across the wrist. Identify these sheaths as a thin fascial covering on the tendons deep to the retinaculum. Completely clean and follow all the tendons across the dorsum of the hand to the digits.

C. Vessels and Nerves

The main vessels in the posterior compartment are the **posterior interosseous artery,** a branch of the common interosseous artery dissected later in the anterior compartment and the **posterior interosseous vein.** Coursing with these vessels in the **posterior interosseous nerve** (deep radial nerve) distal to the supinator. Observe these vessels and nerve running distally by separating the extensor digitorum and extensor carpi ulnaris muscles (p. 391).

D. Dorsum of the Hand

On the dorsum of the hand, review the courses and distributions of all the tendons from the extensor forearm muscles to the digits. Free all tendons from the fascia distal to the extensor retinaculum. Observe the tendons that participate in the formation of the extensor expansions (aponeuroses). Also, review the patterns of cutaneous innervation and positions of the dorsal venous plexus and cephalic and basilic veins. In the floor of the snuff box (bounded by the extensor pollicis longus and brevis and abductor pollicis longus muscles), follow the **radial artery** as it passes laterally and dorsally around the radius and lateral carpal bone (scaphoid) (p. 364). Follow and identify the radial artery to the first palmar interosseous space on the dorsum of the hand, where it pierces the first dorsal interosseous muscle to enter the deep palm to contribute to the deep palmar arch. This deep course will be observed during dissection of the palm. At the level of the distal row of carpal bones, the dorsal carpal branch of the radial artery forms a **dorsal carpal arch** with the dorsal carpal branch of the ulna artery. From the dorsal carpal arch, **dorsal metacarpal arteries** pass distally, contributing to the dorsal digital arteries of the second to fifth digits. The dorsal metacarpal arteries and arch are deep to the long extensor tendons. Review the cutaneous nerves.

18. FLEXOR COMPARTMENT OF THE FOREARM

Complete the removal of superficial fascia from the ventral surface of the forearm and wrist (Figure 3.3). In the ventral forearm, note the cutaneous vessels and nerves (p. 385). Carefully cut and remove the deep antebrachial fascia to expose the ventral flexor muscles. In the flexor compartment of the forearm, there are also superficial and deep muscle groups.

A. Superficial Muscle Group

The five muscles in this group originate from the medial epicondyle and supracondylar ridge of the humerus (pp. 360–361; 388–389). From the lateral to the medial side of the forearm, they are the **pronator teres, flexor carpi radialis, palmaris longus** (often absent), **flexor digitorum superficialis** and **flexor carpi ulnaris.** The tendons of these muscles (except the pronator teres) pass ventrally across the wrist to enter the hand. Their courses and insertions in the hand will be dissected later. Clean and separate the tendons of the superficial muscles in the forearm and follow them to the wrist (pp. 388–389). Review their general origins from the medial epicondyle. The flexor digitorum superficialis forms four tendons at the wrist as it enters the hand. The median nerve innervates all the muscles of the superficial compartment, except the flexor carpi ulnaris, which is innervated by the ulnar nerve. The actions of these muscles will become apparent as they are followed in the hand and should now be reviewed.

B. Deep Muscle Group

There are three muscles in the deep group: the **flexor digitorum profundus, flexor pollicis longus** and **pronator quadratus** (pp. 362; 389). These muscles originate from the radius, ulna and interosseous membrane. Do not yet reflect any of the superficial muscles to observe the deep muscle group. Retraction of the superficial muscles is adequate. Note that the flexor digitorum profundus also forms four tendons at the wrist that are carried into the hand, similar to those of the superficialis. Observe the flexor pollicis longus on the same plane but lateral to the tendons of the flexor digitorum profundus. The pronator quadratus can be seen by retracting the tendons of the flexor digitorum profundus at the wrist. The pronator quadratus and flexor pollicis longus are innervated by the anterior interosseous nerve, a branch of the median nerve. The two medial heads of the flexor digitorum profundus are innervated by the ulnar nerve, whereas the lateral two heads are innervated by the anterior interosseous nerve. Review the actions of these muscles.

C. Vessels and Nerves

Review the structures and relationships within the cubital fossa (pp. 385–387). All the major nerves and vessels in the fossa pass into the forearm. Review the course of the **radial nerve** from the arm into the cubital fossa deep to this muscle, where it divides into the **superficial radial** and **deep radial** nerves (p. 386). The deep radial nerve passes through the supinator to enter the posterior compartment of the forearm. The superficial branch is cutaneous and passes distally through the forearm between the brachioradialis and the flexor pollicis longus accompanied by the **radial artery** on the medial side of the nerve. In the lower third of the forearm, the superficial radial nerve leaves the undersurface of the brachioradialis and passes onto the dorsum of the wrist and hand for cutaneous innervation to the lateral two or three and one-half digits. Review this dorsal course.

The **ulnar nerve** passes from the posterior compartment of the arm dorsal to the medial epicondyle between the two heads of the flexor carpi ulnaris into the forearm. In the anterior compartment of the forearm, the ulnar nerve travels between the flexor carpi ulnaris superficially and the flexor digitorum profundus deeply (p. 387). In the distal two-thirds of the forearm, the nerve is accompanied on its lateral side by the **ulnar artery** to the wrist. Identify the ulnar nerve and vessels in this course through the forearm. In the forearm, the ulnar nerve innervates the medial two heads of the flexor digitorum profundus and flexor carpi ulnaris and provides cutaneous branches to the wrist and the palmar and dorsal surfaces of the hand.

D. Deep Dissection

The muscle reflections that follow will complete the study of the vessels and nerves in the ventral compartment of the forearm (Figures 3.11, 3.12 and 3.13).

First, reflect the **superficial (humeral) head of the pronator teres** at its midpoint (p. 387). Note that the pronator has two heads of origin and observe that the **median nerve** leaves the cubital fossa by passing between the two heads. The **deep (ulnar) head** of the pronator is extremely small and care should be taken to preserve these muscle fibers deep to the median nerve (Figure 3.11). Note that the **ulnar artery** passes deep to both heads of the muscle to reach the medial side of the forearm. The deep head can now be reflected.

Next, review the extensive origin of the **flexor digitorum superficialis** (p. 389). Its lateral origin is from the oblique line on the ventral surface of the radius (Figure 3.12, A-B). Cut this attachment from the radius and retract it medially, leaving its medial (ulnar) origin intact. Note that the **median nerve** and **ulnar artery** pass deep to the upper free margin of the flexor digitorum superficialis to course through the forearm.

Carefully separate the **median nerve** from the deep surface of the reflected flexor digitorum superficialis and note that the median

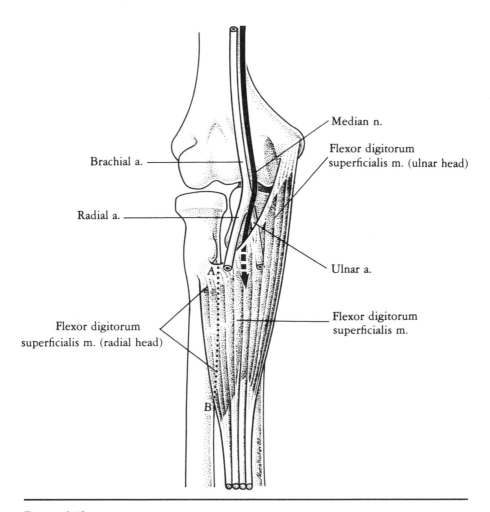

Figure 3.12

nerve runs between the superficialis and the flexor digitorum profundus in the forearm. As the median nerve passes between the two heads of the pronator teres, identify the **anterior interosseous nerve** (p. 387), which branches from the median nerve and passes deeply in the forearm on the interosseous membrane (Figure 3.13). Follow the course of the anterior interosseous nerve in the deep compartment. The anterior interosseous nerve innervates most of the deep muscles of the flexor region and terminates as sensory branches to the wrist and radioulnar joints. The anterior interosseous and median nerves innervate all the muscles of the forearm, except those supplied by the ulnar nerve. Review this pattern of innervation.

The **brachial artery** divides into the **radial** and **ulnar arteries** (pp. 386–387) at the apex of the cubital fossa. The **radial artery**

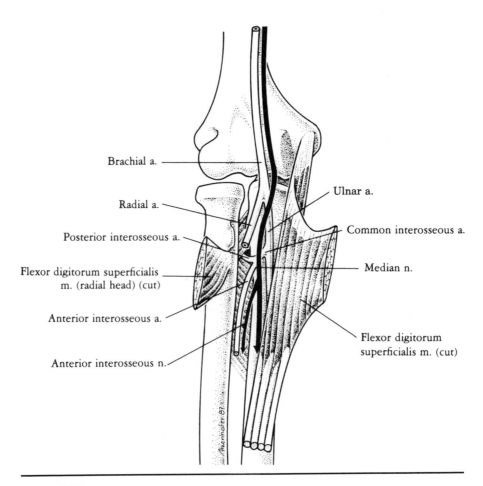

Brachial a.

Radial a.

Posterior interosseous a.

Flexor digitorum superficialis
m. (radial head) (cut)

Anterior interosseous a.

Anterior interosseous n.

Ulnar a.

Common interosseous a.

Median n.

Flexor digitorum
superficialis m. (cut)

Figure 3.13

passes superficial to the upper margin of the flexor digitorum superficialis and then deep to the brachioradialis (Figure 3.12). In the middle third of the forearm, the radial artery accompanies the superficial radial nerve and then distally passes laterally onto the dorsal surface of the hand, as previously seen. Follow the entire course of the radial artery in the forearm. The radial artery gives rise to a small **radial recurrent artery** (p. 387) close to the bifurcation of the brachial artery (Figure 3.11). The radial recurrent artery ascends ventral to the elbow toward the arm deep to the brachioradialis, where it anastomoses with the radial collateral artery to form a collateral circulation at the lateral side of the elbow.

The **ulnar artery** (p. 387) leaves the cubital fossa by passing deep to both heads of the pronator teres to travel deep to the flexor digitorum superficialis (Figure 3.12). Distally on the medial aspect of the forearm, the ulnar artery is deep to the flexor carpi ulnaris. Close to the bifurcation of the brachial artery and deep to the pronator teres, identify the small **anterior** and **posterior ulnar recurrent arteries** from the medial side of the ulnar artery. These ulnar recurrent arteries ascend proximally through the muscles on the medial side of the elbow to anastomose will the **inferior** and **superior ulnar collateral arteries.** These four vessels contribute to the collateral circulation on the medial side of the elbow. At the pronator teres, the ulnar artery is crossed ventrally by the median nerve. At this point branching from the lateral side of the ulnar artery, identify the **common interosseous artery** (p. 387). It is a short trunk and divides at the superior margin of the interosseous membrane into **anterior** and **posterior interosseous arteries** (Figure 3.13). The **posterior interosseous artery** passes superior to the membrane dorsally to enter the posterior compartment of the forearm, where it has been dissected. The posterior artery often gives rise to the interosseous recurrent artery, which anastomoses with the middle collateral artery, a branch of the deep brachial artery of the arm. Identify the **anterior interosseous artery** (with the anterior interosseous nerve), which travels distally on the ventral surface of the interosseous membrane to supply the ventral muscles of the forearm and the wrist, where it sometimes forms a collateral circulation with the hand (p. 362).

19. WRIST (VENTRAL SURFACE)

You will now complete the dissection of the structures on the ventral wrist, which were dissected in the flexor compartment of the forearm (pp. 392–393). Remove the skin that covers the wrist and base of the hand (Figure 3.3). Note that at the wrist, the **radial artery, median nerve** and **ulnar nerve** and **vessels** are superficial and easily damaged. Identify the tendons of the flexor digitorum superficialis and profundus. Clean the superficial tissues and locate the **radial artery** lateral to the **flexor carpi radialis** muscle, the **median nerve** between the **flexor carpi radialis** and the **palmaris longus** and the **ulnar vessels** and **nerve** lateral to the **flexor carpi ulnaris** (Figure 3.14). Before entering the snuff box, the radial artery gives off the palmar and dorsal carpal branches and an inconsistent superficial palmar branch (p. 393) that usually contributes to the superficial palmar arterial arch in the palm (Figure 3.14). The ulnar artery passes on the medial side of the wrist and also forms a palmar carpal branch. The tendons of the flexor digitorum superficialis and profundus and the median nerve pass deep to the flexor retinaculum in the carpal tunnel. The **flexor retinaculum** (pp. 392–393) is a

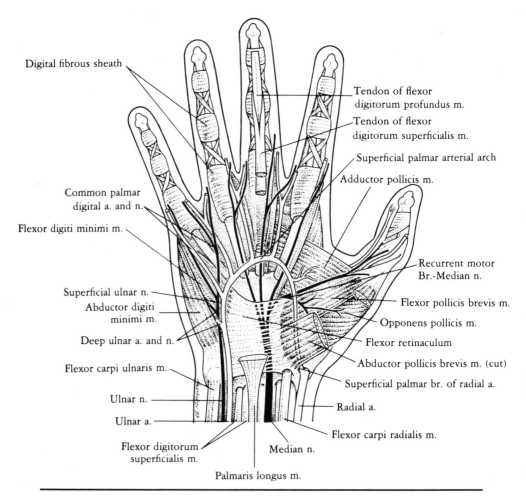

Digital fibrous sheath

Tendon of flexor digitorum profundus m.

Tendon of flexor digitorum superficialis m.

Superficial palmar arterial arch

Adductor pollicis m.

Common palmar digital a. and n.

Flexor digiti minimi m.

Recurrent motor Br.-Median n.

Superficial ulnar n.

Abductor digiti minimi m.

Flexor pollicis brevis m.

Opponens pollicis m.

Deep ulnar a. and n.

Flexor retinaculum

Abductor pollicis brevis m. (cut)

Flexor carpi ulnaris m.

Superficial palmar br. of radial a.

Ulnar n.

Radial a.

Ulnar a.

Flexor carpi radialis m.

Flexor digitorum superficialis m.

Median n.

Palmaris longus m.

Figure 3.14

dense, fibrous arch that keeps the muscles from bowstringing at the wrist during contraction. The retinaculum is attached medially to the hamate and pisiform carpal bones and laterally to the scaphoid and trapezium bones. Clean and identify the proximal and distal margins of the retinaculum and identify the **carpal tunnel** deep to it. The median nerve is the most superficial structure that passes in the carpal tunnel. With scissors, blunt dissect the courses of the palmaris longus and the ulnar nerve and vessels superficial to the retinaculum. Follow the course of the radial artery laterally. Note the insertions of the flexor carpi ulnaris onto the base of the fifth metacarpal bone and of the flexor carpi radialis onto the bases of the second and third metacarpal bones.

20. PALM OF HAND

Complete the removal of skin from the palm and digits, as shown in Figure 3.3. This skin is thick and tightly anchored to the deep fascia of the palm, as indicated by the skin creases. The skin of the palm is tight, allowing the hand to grip firmly. By contrast, the skin of the dorsum of the hand is loose. Deep to the skin, the deep fascia forms a tough **palmar aponeurosis** that covers the central portion of the palm between the thenar and the hypothenar compartments. This aponeurosis receives the insertion of the palmaris longus (p. 392). Reflect the palmaris longus from the aponeurosis and pull it proximally. Note that the palmar aponeurosis sends extensions into the second to fifth digits. Thickening of the aponeurosis forms superficial transverse metacarpal ligaments at the bases of the digits. Remove the palmar aponeurosis with blunt dissection and by cutting with scissors and septa that penetrate deeply and attach to the metacarpal bones. Clean the fascia from the surfaces of the digits. Great care must be taken to protect the vessels and nerves immediately deep to the aponeurosis as it is removed. The fascia that covers the thenar and hypothenar compartments is continuous with the palmar aponeurosis (p. 392) but is thinner.

The palm is divided into four compartments: the **thenar, hypothenar, central** and **adductor-interosseous** compartments. These compartments are separated by fascial layers and septae and provide a well-organized approach to the study and dissection of the hand. Careful dissections of the structures within these compartments will be rewarding to your understanding of the functional anatomy of the hand.

A. Thenar Compartment

Located on the lateral side of the palm, this compartment contains three intrinsic thenar muscles (Figure 3.14) (**abductor pollicis brevis, flexor pollicis brevis** and **opponens pollicis**) and the tendon of the **flexor pollicis longus** (pp. 392–394). The tendon of the flexor pollicis longus is covered by a synovial sheath (radial bursa) that is continuous from the proximal end of the flexor retinaculum to the distal phalanx of the thumb (pp. 362–363). These intrinsic muscles originate from the lateral carpal bones and flexor retinaculum. Their insertions on the proximal and distal phalanges of the thumb should be identified on the skeleton using your textbook and atlas. The three intrinsic thenar muscles are innervated by the (**recurrent) motor branch** (Figure 3.14) of the median nerve (p. 392).

Identify this motor nerve, which branches from the median nerve at the distal end of the flexor retinaculum to enter the thenar muscles. If present, the **superficial palmar branch** of the radial artery enters the palm superficial to or passes through the thenar muscles to contribute to the superficial palmar arch in the central compartment of the hand (p. 393).

B. Hypothenar Compartment

This region contains three intrinsic muscles of the fifth digit that originate from the medial carpal bones and flexor retinaculum (Figure 3.14). They are the **abductor digiti minimi, flexor digiti minimi** and **opponens digiti minimi** (p. 392). Follow and free completely the ulnar nerve and vessels from the fascial tunnel (Guyon's canal) on the anterior surface of the wrist and flexor retinaculum (p. 393). At the midpoint of the retinaculum, identify the division of the ulnar nerve and vessels into the **superficial** and **deep ulnar** nerves and vessels (Figure 3.14) (pp. 393–394). The superficial nerve innervates the **palmaris brevis** and continues to the central compartment, where it becomes cutaneous (**common palmar digital nerves**) to the ulnar one and one-half digits. The deep branches of the ulnar nerve and vessels penetrate between the abductor and the flexor digiti minimi to reach the deeper adductor-interosseous compartment. The deep ulnar nerve innervates the three hypothenar muscles, the third and fourth lumbrical muscles, all the interossei muscles and the adductor pollicis. This deeper course will be dissected in the interosseous compartment.

C. Central Compartment

To follow the structures from the forearm into the central palm and digits, open the carpal tunnel by cutting the flexor retinaculum (pp. 388–389; 392–394). Place a probe deep to the retinaculum and use the scalpel to cut along the probe to avoid injury to the contents of the carpal tunnel. Most superficially in the tunnel is the **median nerve.** Just distal to the flexor retinaculum and deep to the palmar aponeurosis (already removed), review the course of the recurrent motor branch of the median nerve to the muscles of the thenar compartment and identify and follow the **common palmar digital nerves** from the median nerve to supply the lateral three and one-half digits. The digital nerves pass deep to the palmar aponeurosis with the digital vessels. Two small muscular branches of the median nerve are provided to the first two lumbrical muscles. Next, follow the tendons of the **flexor digitorum superficialis** and **flexor digitorum profundus** through the carpal tunnel into the palm. In the carpal tunnel, the four tendons of the flexor digitorum profundus lie deeply in a single row. The four tendons of the flexor digitorum superficialis are arranged with the third and fourth digits superficial to the tendons of the second and fifth digits. In the central compartment, they fan distally toward the second to fifth digits in two planes. A common synovial sheath (ulnar bursa) follows the superficial and deep long tendons into the midpalm (pp. 362–363). Only the synovial sheaths that cover the two tendons of the fifth digit are continuous to the distal phalanx. Identify the four **lumbrical** muscles, noting that they have a moving point of origin from the four tendons of the flexor digitorum profundus in the palm (p. 361). The lumbrical muscles continue into the second to fifth digits and insert onto the radial side of the extensor expansion on the dorsum of the digits. The lateral two lumbrical muscles are innervated by the median nerve, whereas the medial two lumbrical muscles are innervated by the ulnar nerve.

Superficial in the central compartment at the horizontal plane of the extended thumb is the first of the two arterial arches of the palm. This **superficial palmar arterial arch** (Figure 3.14) is deep to the palmar aponeurosis and is formed primarily by the **superficial branch of the ulnar artery** from the medial side of the hand (p. 393). A lateral contribution is often provided by the superficial palmar branch of the radial artery. The superficial arch curves across the palm, providing common palmar digital arteries superficial to the flexor tendons. These digital arteries pass distally to the webs of the fingers, where they join the palmar and dorsal metacarpal arteries to form the proper palmar digital arteries along the margins of the second to fifth digits (p. 393).

Follow the superficial and deep long flexor tendons into the digits. Note that the tendons enter a **digital fibrous sheath** that forms a fibrous tunnel (Figure 3.14) on the ventral side of the second to fifth digits (pp. 392–393). Similar in function to the flexor retinaculum, the fibrous sheaths prevent bowstringing of the long flexor tendons by keeping them close to the surface of the phalanges. With scissors, open the sheaths longitudinally along each digit (p. 364). Within the fibrous sheath, elevate the tendons and note the unique arrangement of the superficial digital tendons, which split to allow passage of the tendons of the flexor digitorum profundus to reach the distal phalanx (Figure 3.14). Synovial sheaths are found on the long tendons as they pass deep to the digital fibrous sheaths. The long tendons are connected by the **vincula brevia** and **longa** (p. 364) to the digital fibrous sheaths. These sheaths are folds of membranes that conduct blood vessels to the tendons.

D. Interosseous-Adductor Compartment

The deepest compartment of the palm contains the **palmar** and **dorsal interosseous** muscles, **adductor pollicis** muscle and **deep branch of the ulnar nerve** and **deep palmar arterial arch** (pp. 364; 394). It is best to follow the dissections of this compartment on only one side of the body. To approach these deep structures, it is necessary to make a transverse cut through the fleshy part of the flexor digitorum superficialis immediately superior to the wrist and tendons of the flexor digitorum profundus in the distal carpal canal proximal to the origins of the lumbrical muscles. The cut ends of these muscles can be pulled proximally and distally from the carpal canal to provide access to the deep palm. Cut the center of the superficial palmar arterial arch and reflect it to the sides of the hand. The **interosseous** muscles are wedged between the metacarpal bones. Clean the fascia from the palmar surface of the interosseous muscles toward the webs of the fingers. The **palmar** and **dorsal interosseous** muscles can be identified best by following the interosseous tendons into the digits and observing the side of insertion (radial or ulnar) of the interosseous tendons into the **extensor expansion** of the second to fifth digits (pp. 364–365). The **deep transverse metacarpal ligaments** bind and hold the heads of the metacarpal bones. Note that these ligaments separate the lumbrical muscles from the interosseous muscles as these muscles pass into the digits. Cutting the transverse ligament makes it easier to follow these tendons to their insertions on the extensor expansion. Complete the details of the dissection of the extensor expansion. Identify its **central** and **lateral bands** on several digits and understand their formation on the dorsum of the digits. Understand the pattern of insertion of all muscles into the expansion and describe their actions.

Identify the **adductor pollicis** with its **oblique head** of origin from the capitate bone and its **transverse head** from the shaft of the third metacarpal bone (p. 364). The adductor pollicis covers the lateral interosseous muscles and inserts on the proximal phalanx of the thumb.

The **deep branches of the ulnar artery** and **nerve** penetrate the base of the hypothenar muscles to enter the interosseous-adductor compartment (p. 394). Follow the deep branches from the hypothenar compartment and identify them in the fascia on the ventral surface of the interosseous muscles in the deep compartment. Laterally, they pass deep to the adductor pollicis. The deep ulnar nerve innervates the hypothenar muscles, all the interosseous muscles, the adductor pollicis and the third and fourth lumbrical muscles. The deep ulnar artery forms the deep palmar arterial arch (p. 394) by uniting with the deep course of the radial artery after it passes through the first dorsal interosseous muscle. The arch is distal to the deep ulnar nerve. From the deep arterial arch, identify the **palmar metacarpal arteries,** which course distally on the surfaces of the interosseous muscles to meet the common palmar digital (from the superficial arch) and dorsal metacarpal arteries (from the dorsal carpal arch) to form the proper digital arteries of the digits. As the radial artery passes through the first dorsal interosseous muscle, it provides a princeps pollicis artery to the thumb and a radialis indicis artery to the index finger.

21. ARTICULATIONS

For courses that include scheduled dissections of the joints, brief descriptions will be provided for the shoulder, elbow and wrist joints. These joints should be dissected on only one limb and the articulating surfaces of the bones should first be thoroughly reviewed.

A. Shoulder Joint

The shoulder, or glenohumeral, joint (p. 350) is a freely movable ball-and-socket articulation. It is prone to dislocations because the socket, formed by the glenoid cavity of the scapula and the glenoid labrum, is extremely shallow. Stability is provided primarily by ligaments and surrounding muscles.

To display the joint posteriorly, identify and cut the insertions of the supraspinatus, infraspinatus and teres minor. Several bursae are associated with these muscles. If the acromion was not cut previously, saw it off at its base. The capsule of the joint attaches to the glenoid labrum superiorly and to the head of the humerus distally (p. 350). Posteriorly, make a vertical slit in the capsule. Identify the **glenoid cavity, glenoid labrum** and **head of the humerus.** Anteriorly, the capsule is strengthened by three glenohumeral ligaments (**superior, middle** and **inferior**) and the **tendon of the long head of the biceps,** all of which are seen from this posterior view. Define and identify these ligaments in the anterior wall of the capsule, noting that several of them may be indistinct and fused together. It is often necessary to chisel away part of the posterior aspect of the head of the humerus at the anatomic neck to observe these ligaments. In addition, anteriorly identify the coracoacromial and coracohumeral ligaments, which are accessory supports for the shoulder joint. Observe the relationship of the long head of the biceps to the joint and review the insertions of the subscapularis, teres major, latissimus dorsi and pectoralis major.

B. Elbow

Expose the capsule of the elbow joint by stripping away the muscles on the medial and lateral aspects of the articulation (p. 351). The elbow joint is a hinge between the upper ends of the radius and ulna and the distal humerus. On the skeleton, identify the **trochlea** and **capitulum** of the distal humerus, **trochlear notch** on the ulna and **head of the radius.** The capsule attaches to the margins of the humerus above and to the radius and ulna below, enclosing the proximal radioulnar joint as well. The capsule is thickened medially to form the **ulnar collateral ligament,** which passes between the medial epicondyle and the ulna. Laterally, the capsule is thickened to form the **radial collateral ligament** between the lateral epicondyle and the **annular ligament.** The annular ligament is the strong, circular ligament that provides the pivoting action between the head of the radius and the radial notch of the ulna.

C. Distal Radioulnar and Wrist Joints

Similarly, these two articulations are closely related to the distal end of the forearm. At the distal radioulnar joint, the **head** of the ulna articulates with the **ulnar notch** of the radius (p. 351). The wrist joint (radiocarpal) is formed primarily by the articulating surfaces of the radius and **articular disc** proximally with the scaphoid, lunate and triquetral carpal bones distally. Strip away the muscle at the wrist and observe that the capsule covers both of these joints and is strengthened by **ulnar** and **radial collateral ligaments.**

CHAPTER FOUR
LOWER LIMB

22. SUPERFICIAL STRUCTURES

The lower limb functions primarily in support and locomotion; thus, most of its structures are modified for strength rather than for grasping and mobility, as in the upper limb. In addition, the articulations between the pelvic girdle (formed by the os coxae) and the axial skeleton of the lower limb are more rigid than those of the pectoral girdle in the upper limb. This greater rigidity is demonstrated by the fused attachments of the sacrum and os coxae (sacroiliac joint) and by the stability provided by the ball-and-socket hip joint.

On the skeleton, review and identify on the os coxae and femur (pp. 396–402) the **iliac crest, anterior superior** and **inferior iliac spines, symphysis pubis, ischiopubic ramus, ischial tuberosity, lateral (outer) aspects of the sacrum and iliac bones, greater** and **lesser trochanters** and **intertrochanteric crest** and **line.** Palpate the **greater trochanter,** which is a hand-breadth below the **tubercle of the iliac crest.** Project to the surface of the body the positions of the **sacrotuberous** and **sacrospinous ligaments** and review the formations of the **greater** and **lesser sciatic foramina.**

At the knee, identify on the skeleton the **patella, medial** and **lateral femoral condyles** and **tibial condyles** (p. 405). The **tibial tuberosity** is found inferior to the patella on the anterior surface of the tibia. At the ankle, the **lateral** and **medial malleoli** can be seen on the skeleton. Also, identify the **calcaneus, navicular,** three **cuneiform** and **talus** tarsal bones of the foot (pp. 406–407). Also, note the **metatarsal** and **phalangeal** bones.

23. GLUTEAL REGION

The gluteal region is the area of the buttocks that contains three large gluteal and several small muscles. Vessels and nerves in this area are derived from the internal iliac vessels and sacral plexus within the pelvis. These vessels and nerves enter the gluteal compartment via the **greater** and **lesser sciatic foramina.** Define the boundaries of these foramina and list their contents. Review the bony parts of the lateral aspects of the os coxae and superior end of the femur before dissecting this area.

Turn the body to the prone position and make the skin incisions on the gluteal area and posterior thigh to a point just inferior to the knee, as illustrated in Figure 4.1. Deep to the skin, note the dense nature of the fat and fascia that cover the gluteal region. Note the positions of the sacrotuberous and sacrospinous ligaments, ischial tuberosity and greater trochanter.

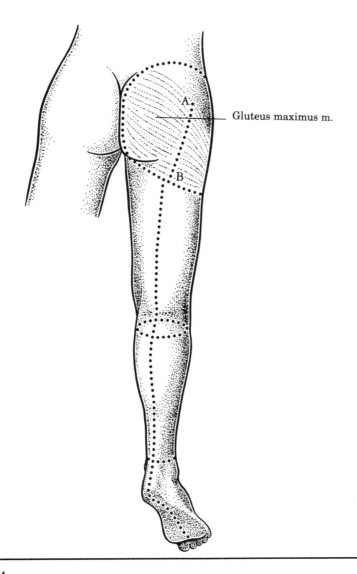

Gluteus maximus m.

Figure 4.1

A. Muscles

Remove the superficial fascia from the gluteal region and identify the rhomboid-shaped **gluteus maximus** muscle (pp. 418–419; 440). Define clearly its superior and inferior borders (Figure 4.2) and note its origins from the dorsum of the sacrum, ilium and sacrotuberous ligament. Identify and cut the **gluteal aponeurosis** along the superior border of the muscle. This aponeurosis provides part of the origin of the gluteus maximus from the iliac crest and overlies part of

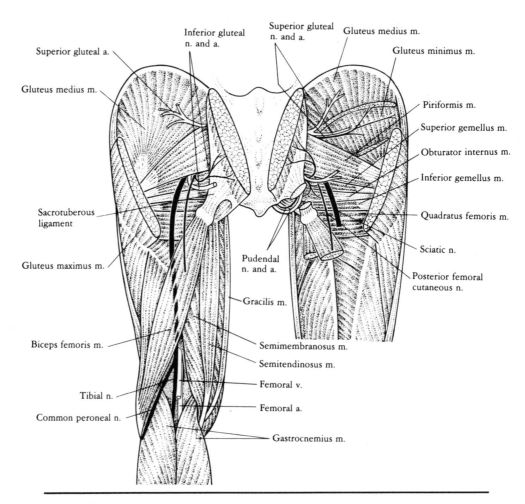

Figure 4.2

the gluteus medius. Follow the fibers of the gluteus maximus later-ally and inferiorly toward their insertions on the **gluteal tuberosity** and **iliotibial tract** (pp. 418–419). Expose and separate the superior structures and fascia of the posterior thigh to observe the inferior border and insertion of the gluteus maximus and the courses of other contents of the gluteal region. Be sure the *entire* inferior border of the muscle is completely cleaned and exposed. Cut and reflect the muscle along the line indicated in Figure 4.1, A-B. Reflect the medial and lateral ends of the muscle. Medially, elevate the inferior border of the muscle and shave with the scalpel the origin of the gluteus maximus from the coccyx, sacrum and **sacrotuberous ligament,** which attaches between the ischial tuberosity and the sacrum (Fig-ure 4.2). Completely cut the fibers that originate from the sacrotuberous ligament as the muscle is pulled medially. On the deep surface of the gluteus maximus, identify the **superior gluteal vessels,** which are attached to its superior aspect, and the **inferior gluteal vessels** and **nerve,** which are attached to its inferior aspect (Figure 4.2). Follow and clean these vessels and nerve. To reflect the gluteus maximus completely, cut out a cube of the inferior fibers where the inferior vessels and nerve enter the muscle. Review the actions, attachments and innervation of this muscle in the textbook.

With the gluteus maximus reflected, identify the **gluteus medius** muscle deep and superior to the gluteus maximus (pp. 418; 441–443). Elevate the inferior border of the gluteus medius and identify the fascial plane deep to it. Incise a part of the gluteus medius (Figure 4.2) to expose the deeper **gluteus minimus** muscle (p. 441). Note the courses of the **superior gluteal vessels** and **nerve** in the plane between the gluteus medius and the gluteus minimus. Review the origins, insertions, actions and innervations of the three gluteal muscles. Identify the **tensor fasciae latae** muscle (pp. 416–417; 438) in the fascia lateral to the gluteus medius. The tensor muscle is innervated by the superior gluteal nerve.

On the same plane, but inferior to the gluteus medius, observe and clean the superior and inferior borders of the **piriformis** muscle (pp. 420; 441), which enters the gluteal compartment through the greater sciatic foramen. The piriformis can be followed to its insertion on the greater trochanter. Inferior to the piriformis, identify the **obturator internus** muscle and tendon making a 90-degree turn through the lesser sciatic foramen to pass dorsal to the hip joint and insert on the greater trochanter. The latter course of the tendon is covered by the two small **superior** and **inferior gemelli** (p. 420) muscles (Figure 4.2), which insert with the obturator internus. Identify the gemelli at the superior and inferior borders of the tendon of the obturator internus, which can be seen embedded between the gemelli. Inferior to these three muscles, not the large **quadratus femoris,** which extends between the **ischium** and the **quadrate tubercle of the femur.** Review the attachments, actions and innervations of this group of muscles. The obturator internus and superior gemellus are innervated by the nerve to the obturator internus and the inferior gemellus and quadratus femoris are innervated by the nerve to the quadratus femoris.

B. Vessels and Nerves

Most of the vessels and nerves enter the gluteal area via the greater and lesser sciatic foramina (pp. 440–443). The piriformis is the key landmark in this area. Superior to this muscle, review the courses of the **superior gluteal vessels** and **nerves.** These vessels and nerves enter this area through the greater notch and lie superior to the piriformis in the fascial plane between the gluteus medius and the gluteus minimus. The superior gluteal nerve supplies the gluteus medius, gluteus minimus and tensor fasciae latae.

Inferior to the piriformis, identify the **inferior gluteal vessels** and **nerve,** which innervates the gluteus maximus. The inferior gluteal nerve is the sole supply of the gluteus maximus. The nerve to the quadratus femoris passes on the bone deep to the obturator internus and superior gemellus to innervate the quadratus femoris and inferior gemellus.

The largest nerve in this area is the **sciatic nerve** (L4 to S3), which is composed of two nerves: the tibial and common peroneal. Observe that the sciatic nerve passes inferior to the piriformis (Figure 4.2) and then crosses the obturator internus, the two gemelli and the quadratus femoris dorsally (pp. 442–443). The sciatic nerve is the main motor and sensory nerve supply to the posterior thigh, leg and foot. Inferior to the quadratus femoris, the sciatic nerve enters the posterior compartment of the thigh. The **posterior femoral cutaneous nerve** descends on the medial side of the sciatic nerve. Follow the sciatic nerve and posterior femoral cutaneous nerve a short distance into the posterior thigh.

Finally, in the gluteal region, the intermediate courses of the **pudendal nerve** and **internal pudendal artery** and **vein** (pp. 441; 443) will be followed. To identify these vessels and nerve, locate and incise the sacrotuberous ligament at its midpoint (Figure 4.2). This maneuver exposes the vessels and nerves that cross the **ischial spine** and **sacrospinous ligament.** The pudendal nerve is the most medial structure that crosses the spine dorsally to enter the pudendal canal to reach the perineum. Lateral to this nerve, the internal pudendal vessels cross the ischial spine to reach the perineum. The **nerve to the obturator internus** and superior gemellus lies lateral to the vessels. Identify the sacrospinous ligament deep to the pudendal nerve and vessels.

24. ANTERIOR (EXTENSOR) COMPARTMENT OF THE THIGH

With the body in the supine position, make skin incisions on the anterior thigh to a point just inferior to the knee, as indicated in Figure 4.3. Be careful that the superior incision is 2.5 cm below and parallel to the inguinal ligament. Remove the skin and superficial fascia separately to expose the dense, deep fascia of the thigh, the

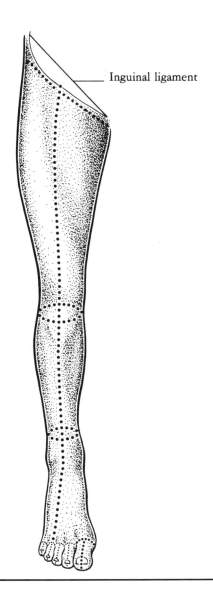

Inguinal ligament

Figure 4.3

fascia lata. Within the superficial fascia, many cutaneous nerves (mostly originating from the femoral nerve) and veins are found (pp. 436–437). Identify the upper course of the **greater saphenous vein** as it enters the thigh after passing posterior to the medial condyle at the knee. Follow its course to the saphenous opening in the fascia lata, where it passes through this opening and terminates in the deeper femoral vein. The greater saphenous vein receives many tributaries, including the superficial circumflex iliac, superficial epigastric and external pudendal veins, as it enters the saphenous opening. The superficial epigastric vein communicates through the abdominal wall with veins of the thorax and axilla to form a cutaneous collateral venous connection between the two venae cavae. This venous flow can enlarge during conditions that block venous return to the inferior vena cava, thus providing an alternate route of venous flow to the heart.

The **saphenous opening** (pp. 198; 431; 437) is a slit in the fascia lata just inferior to the medial attachment of the inguinal ligament to the pubic tubercle. Note that it is oval and has superficial and deep laminae. The opening, filled with cribriform fascia and perforated by the greater saphenous vein, is the site of femoral hernias. These hernias pass through the femoral canal and bulge the skin over the opening.

Note that there are many cutaneous nerves (p. 437) that supply segments of the skin. For the most part, these nerves are derived from the femoral nerve (lateral, intermediate and medial) or the obturator nerve. Branches of the ilioinguinal and genitofemoral nerves (branches of the lumbar plexus) also overlap the upper aspect of the thigh. Preserve any of these nerves as you proceed.

Deep to the superficial fascia, observe that the **fascia lata** forms a tough layer that encloses the structures of the thigh like a stocking. Superiorly, it is attached to the inguinal ligament, iliac crest, sacrotuberous ligament, ischial tuberosity, ischiopubic ramus and symphysis pubis. Laterally, it is thickened to form the **iliotibial tract,** which receives part of the insertions of the gluteus maximus and tensor fasciae latae and attaches inferiorly to the lateral tibial condyle. Identify this tract on the lateral surface of the thigh. Lateral, medial and intermediate intermuscular septa pass deeply from the fascia lata to attach to the femur, forming extensor, flexor and adductor compartments. Note these compartments on a cross section of the thigh. At the knee, the fascia thickens to form the lateral and medial retinacula of the patella and then continues as crural fascia in the leg.

With scissors, carefully excise and remove the fascia lata with an anterior, vertical cut, thus opening the anterior compartment. This compartment is separated medially from the adductor (medial) compartment by the medial intermuscular septum and is separated laterally from the flexor (posterior) compartment by the lateral septum. The main contents of the anterior extensor compartment are the **quadriceps femoris** and **sartorius** muscles and the distributions of the **femoral vessels** and **nerve** (Figures 4.4 and 4.5). The upper aspect of the anterior compartment at the junction between the thigh

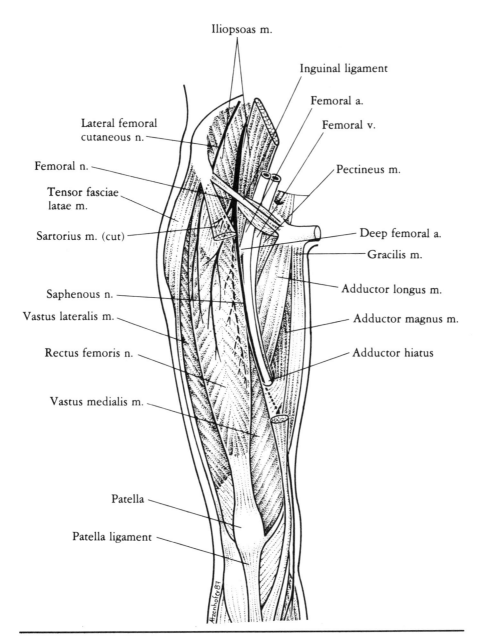

Iliopsoas m.

Inguinal ligament

Femoral a.

Femoral v.

Lateral femoral
cutaneous n.

Femoral n.

Tensor fasciae
latae m.

Pectineus m.

Sartorius m. (cut)

Deep femoral a.

Gracilis m.

Saphenous n.

Adductor longus m.

Vastus lateralis m.

Adductor magnus m.

Rectus femoris n.

Adductor hiatus

Vastus medialis m.

Patella

Patella ligament

Figure 4.4

and the lower trunk is called the **femoral triangle** (pp. 416; 438). Note that it is bounded medially by the medial border of the **adductor longus,** laterally by the medial border of the **sartorius** and superiorly by the inguinal ligament. In the floor of the femoral triangle are the iliopsoas, pectineus and adductor longus muscles.

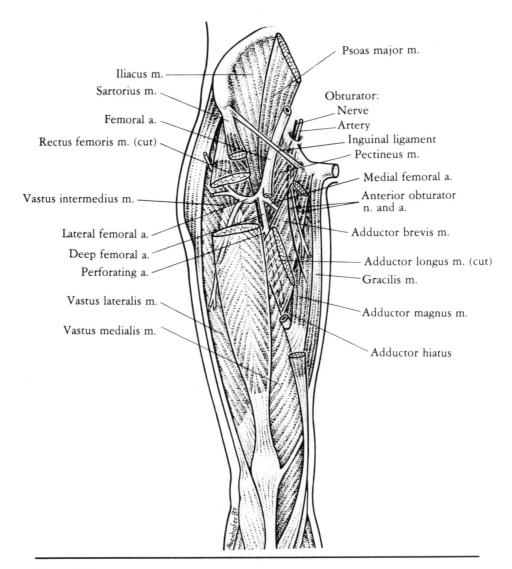

Figure 4.5

The latter two muscles are structures of the adductor compartment. The main trunks of the femoral vessels and nerve enter the thigh first within the femoral triangle.

Review the courses of the femoral nerve and external iliac artery and vein as they pass deep to the inguinal ligament to enter the anterior thigh (pp. 430; 438–439). The names of the external iliac vessels change to femoral as they enter the femoral triangle. Here, the **femoral nerve** is lateral and the **femoral vein** is medial to the

femoral artery (Figure 4.4). Isolate the main trunks of these vessels and nerve by removing the dense fascia in the upper part of the triangle. As the two vessels pass deep to the inguinal ligament, they are covered by the **femoral sheath,** which is primarily a continuation of the transverse fascia along the vessels. Divisions of this sheath form a lateral compartment for the femoral artery, a middle compartment for the vein and a medial compartment, the femoral canal, which contains lymph nodes and fat, that opens at the saphenous opening. Elevate the femoral nerve and remove the fascia deep to it to identify the **iliopsoas** muscle (Figure 4.5). Elevate the femoral vessels and remove the fascia deep to them to identify the **pectineus** muscle. These vessels and nerve will be dissected in detail later.

A. Muscles

Begin your study by reviewing the major parts of the femur related to the thigh (p. 403). Superiorly, identify the **greater** and **lesser trochanters, trochanteric fossa, head** and **neck of the femur, intertrochanteric line** and **crest, quadrate tubercle, gluteal tuberosity, pectineal line** and **linea aspera.** Inferiorly, observe the **lateral** and **medial condyles, lateral** and **medial epicondyles, adductor tubercle** and **intercondylar fossa.**

With the fascia lata removed (except for the iliotibial tract), identify the more superficial **sartorius** muscle (p. 438). Clean this muscle (Figure 4.5) from its fascia and note that it crosses the thigh obliquely from its origin on the anterior superior iliac spine to its insertion on the medial surface of the tibia. Its insertion (**pes anserinus**) is in common with the **gracilis** and **semitendinosus** muscles on the medial surface of the tibia. Describe the actions of the sartorius and identify its innervation by the femoral nerve.

Next, identify the **quadriceps femoris** muscle group (pp. 416; 438–439). This muscle group is the main anterior mass that extends the knee (Figures 4.4 and 4.5). It is divided into four muscles:

1. The **rectus femoris** arises mainly by a straight head from the anterior inferior iliac spine. A reflected head also arises superior to the acetabulum of the hip joint. Follow the rectus femoris through the middle of the thigh to its insertion on the patella.

2. The **vastus lateralis** forms the large muscle mass on the lateral side of the thigh. At the knee, this muscle is palpable on the lateral side of the patella, where it inserts.

3. The **vastus medialis** is the most medial part. Follow the fibers

of this muscle to their attachments on the patella and note that the lowest fibers are almost horizontal.

4. The **vastus intermedius** arises more deeply from the femur. It descends deep to the rectus femoris and between the other two vastus muscles to insert on the patella. To expose this muscle, reflect the rectus femoris.

From their insertions on the patella, identify the final insertion of the quadriceps via the **patellar ligament** to the **tibial tuberosity.** The quadriceps is the primary extensor muscle of the knee and is innervated by the femoral nerve.

Identify the **iliopsoas** muscle (p. 416) deep to the femoral nerve in the femoral triangle, as described previously. It is formed by the fused **iliacus** and **psoas major** (p. 416). Remove the fascia around and deep to the femoral nerve to observe the iliopsoas. The iliopsoas originates from the lateral pelvic and posterior abdominal walls. Follow its fused fibers deep to the femoral nerve to their insertions on the lesser trochanter in the floor of the femoral triangle. The iliopsoas is a strong flexor muscle of the hip joint (or of the trunk with the lower limb stationary).

Next, in the extensor compartment, identify the **adductor canal** (pp. 416–417), which is a fascial and muscular passage through the anterior thigh. The canal begins at the midpoint of the thigh, where the **sartorius** crosses the **adductor longus.** Cut and reflect the sartorius at its midpoint to dissect the canal. The femoral vessels pass deep to the sartorius to enter the canal along with the **saphenous (cutaneous) nerve** and **nerve to the vastus medialis.** The canal passes obliquely and deep to the sartorius, which forms its roof. The canal is bounded medially by the **adductor longus** and **magnus** and laterally by the **vastus medialis.** Distally, the canal ends at the **adductor hiatus,** a separation in the insertion of the adductor magnus at the **adductor tubercle.** Separate and follow the femoral artery and vein through the adductor canal as well as the two branches of the femoral nerve listed previously.

B. Nerves

The **femoral nerve** (L2 to L4) is the primary motor and cutaneous innervation to the anterior thigh. It is the largest nerve of the lumbar plexus (pp. 416–417). Identify the femoral nerve as it passes deep to the inguinal ligament lateral to the femoral artery into the femoral triangle (Figure 4.4). The femoral nerve lies in the groove between the psoas and the iliacus. Many of its branches to the sartorius, quadriceps femoris and part of the pectineus can be observed and followed. Note that there are many cutaneous branches of the femoral nerve. One of the largest cutaneous branches is the saphenous nerve, which was observed passing through the adductor canal and then running distally with the greater saphenous vein on the medial side of the leg. The femoral nerve also supplies branches to both hip and knee joints.

C. Vessels

The **femoral artery** (pp. 416–417; 430) enters the femoral triangle within the femoral sheath as a continuation of the external iliac artery deep to the inguinal ligament (Figure 4.5). This artery lies between the femoral nerve laterally and the femoral vein medially. Clean the fascia and sheath completely from the entire course of the artery and follow the artery as it descends through the triangle and enters the adductor canal at the apex of the triangle. The femoral artery then descends through the adductor canal to enter the popliteal fossa by passing through the adductor hiatus. Identify and separate the course of the femoral vein from the artery. Be aware that the fascia is dense around the superior course of the femoral artery and vein and has to be removed carefully to allow identification of the branches of the artery and vein.

Just inferior to the inguinal ligament are three small, subcutaneous vessels that arise from the femoral artery (superficial epigastric, superficial circumflex iliac and superficial external pudendal arteries). These vessels are extremely small and are not critical to dissect.

In the femoral triangle, identify the **deep femoral artery** (profunda femoris) (Figure 4.5), which arises from the deep lateral aspect of the femoral artery (p. 439). Clean and identify its descending course posterior and deep to the femoral artery and then deep to the adductor longus; the femoral artery is ventral to this muscle. Elevate the femoral artery to clean the entire course of the profundus artery. The dense fascia around the deep femoral vessels has to be removed. The continuation of the deep femoral artery posterior to the adductor longus gives rise to three or four **perforating arteries.** These arteries penetrate dorsally through the adductor magnus to enter the posterior compartment of the thigh, where they provide the major blood supply. Note that the courses of the femoral and deep femoral veins parallel the courses of the arteries and the veins should be separated from the arteries. The femoral vein often has to be removed to expose the arterial distribution.

From the deep femoral artery close to its origin in the femoral triangle, identify the **medial** and **lateral circumflex femoral ar-**

teries. Clean and follow the medial femoral circumflex artery as it passes deeply between the pectineus and the iliopsoas in the floor of the triangle. Most of its blood is distributed to posterior thigh and adductor muscles, with important branches to the hip joint. The lateral femoral circumflex artery branches from the lateral side of the deep femoral artery. It passes deep to the sartorius and rectus femoris, where it then divides into **ascending, transverse** and **descending branches** (Figure 4.5). The ascending artery passes deep to the tensor fasciae latae, where it anastomoses with the superior gluteal artery of the hip. The transverse branch passes deep to the vastus lateralis and encircles the neck of the femur to meet the medial circumflex femoral, inferior gluteal and perforating arteries, forming the cruciate anastomosis. The descending branch can be followed inferiorly on the surface of the vastus lateralis.

25. MEDIAL (ADDUCTOR) COMPARTMENT OF THE THIGH

This compartment includes muscles that arise from the external surface of the pubis and ischiopubic ramus. These muscles are innervated by the obturator nerve. Some of them have already been described because they are related to the femoral triangle and adductor canal (pp. 416–417; 438–439). Review the attachments, actions and innervations of these muscles in your textbook.

First, identify the **gracilis,** which is a long, vertical muscle on the medial aspect of the thigh. It inserts with the sartorius and semitendinosus at the pes anserinus. The remaining muscles form three layers (Figures 4.4 and 4.5):

1. In the superficial layer, identify the **pectineus** and **adductor longus.** Identify their borders and review their relationships to the femoral triangle and adductor canal. Note that the pectineus is usually innervated by the femoral nerve. Cut and reflect the adductor longus.

2. In the middle plane, observe the **adductor brevis.** It is necessary to cut and reflect the adductor longus at its midpoint to observe the adductor brevis. Identify the anterior and posterior branches of the **obturator nerve** on the anterior and posterior surfaces of the adductor brevis.

3. In the deepest layer, the **adductor magnus,** the largest muscle of the adductor group, can be observed. It has perforations for passage of the perforating arteries and splits to form the **adductor hiatus** distally for passage of the femoral vessels to the popliteal

fossa at the posterior aspect of the knee. The adductor magnus has transverse, oblique and vertical fibers. Its vertical fibers insert on the **adductor tubercle** of the femur.

An additional muscle of this compartment is the **obturator externus** (p. 417). It arises from the external surface of the obturator membrane and passes deep to the other adductor muscles to reach the trochanteric fossa. Identify this muscle on one side by reflecting the pectineus.

The **obturator nerve** (p. 439) is formed by anterior fibers of the lumbar plexus (L2 to L4). It leaves the pelvis through the obturator canal. Identify the anterior and posterior divisions of the obturator nerve, which pass around the adductor brevis (Figure 4.5). This nerve supplies fibers to the hip joint and branches to the adductor muscles, except for the pectineus, which is usually innervated by the femoral nerve. An accessory obturator nerve is sometimes present.

The **obturator artery** branches from the internal iliac artery and passes from the pelvis through the obturator canal to enter the medial region of the thigh. Anterior and posterior branches are usually extremely small and lie on the ventral and dorsal surfaces of the adductor brevis. These branches supply most of the muscles of the medial thigh and hip joint. The vein parallels the artery.

26. POSTERIOR (FLEXOR) COMPARTMENT OF THE THIGH

This compartment contains the hamstring muscles, which originate on the ischial tuberosity and insert on the tibia or fibula. They function in extending the hip and flexing the knee. Review the bony aspects of the distal femur and proximal tibia and fibula. Complete the removal of the superficial and deep fasciae from the posterior thigh to identify the muscles and other structures (Figure 4.2). Develop the continuity of structures that pass between the gluteal region and the posterior thigh. Two muscles of the posterior thigh, the **semitendinosus** and **semimembranosus,** descend on the medial side of the thigh and knee. The third muscle of this group, the **biceps femoris,** descends on the lateral aspect of the thigh and knee (pp. 419–420; 442–443). Review the origins and insertions of these muscles. Note that the **biceps femoris** has a **long** and a **short head.** The long head arises from the ischial tuberosity and crosses the sciatic nerve to reach the lateral side of the thigh. At the level of the mid-thigh, the long head is joined deeply by the short head. The biceps femoris passes the knee laterally to insert on the head of the fibula. Follow the **semimembranosus** from the ischial tuberosity to the medial side of the knee, where it inserts almost horizontally on the medial condyle of the tibia. During most of its course, the semimembranosus is broad and fleshy. The **semitendinosus** lies on the dorsal surface of the semimembranosus and becomes a narrow tendon in the inferior third of the thigh (pp. 420; 442–443). The semitendinosus inserts on the medial aspect of the tibia in common with the gracilis and sartorius (pes anserinus). All these hamstring muscles are innervated by the tibial nerve, except for the short head of the biceps femoris, which is innervated by the common peroneal nerve.

Follow the sciatic nerve from the gluteal compartment through the posterior thigh and note that it divides into the **tibial** and **common peroneal nerves** at the superior aspect of the popliteal fossa (pp. 442–443). Observe the course of the common peroneal nerve as it follows the tendon of the biceps femoris around the neck of the fibula to reach the lateral side of the leg. The tibial nerve remains in the midline of the popliteal fossa. The arterial supply of the posterior compartment is primarily from perforating branches of the deep femoral artery. Review the origins of these perforating branches from the deep femoral artery in the anterior compartment and their courses through the adductor magnus to reach the posterior compartment.

27. POPLITEAL FOSSA

This diamond-shaped fossa (pp. 442–443; 445–446) at the posterior aspect of the knee contains the major nerves and vessels that course between the thigh and the leg (Figure 4.2). The superior boundaries are formed by the biceps femoris laterally and the semimembranosus and semitendinosus medially. Inferior to these muscles, identify the **medial** and **lateral heads** of the origins of the **gastrocnemius** muscle, which form the inferior medial and lateral boundaries of the fossa. Complete the removal of the superficial fascia, being careful to identify the terminal course of the **lesser saphenous vein** and the **sural cutaneous nerves** within the fascia. Next, remove the deep fat and fascia within the fossa. The **popliteal artery** and **vein** can be seen passing through the center of the fossa. These vessels are continuations of the femoral artery and vein, which course through the adductor hiatus from the adductor canal of the thigh to reach the popliteal fossa. Review the courses of these vessels in the anterior thigh. The popliteal artery is most ventral, crossing the floor of the fossa and the posterior aspect of the knee joint. The vein is posterior to the artery and receives the lesser saphenous vein within the fossa. The tibial nerve is dorsolateral to the vein and is described in the text that follows. Branches from each of the popliteal vessels include the superior, middle and inferior geniculate arteries and veins and the sural arteries and veins to the gastrocnemius and soleus. It is not necessary to identify each of these vessels. The geniculate branches form an important collateral circulation around the knee with the femoral vessels of the thigh and the branches of the anterior and posterior tibial vessels of the leg. The middle geniculate artery is an important blood supply to the knee joint.

The sciatic nerve usually divides into the **tibial** and **common peroneal** nerves in the superior aspect of the popliteal fossa. Follow the tibial nerve as it descends through the center of the fossa posterior and lateral to the popliteal vein to enter the posterior compartment of the leg deep to the gastrocnemius. In the fossa, the tibial nerve supplies motor branches to the gastrocnemius, soleus, plantaris and popliteus, articular branches to the knee joint and medial sural cutaneous branches to the leg. The common peroneal nerve follows the course of the biceps femoris, giving rise to lateral sural cutaneous branches to the leg and motor branches to muscles of the anterior and lateral compartments of the leg, with its only innervation in the thigh being to the short head of the biceps.

28. LEG

The leg is the segment of the lower limb between the knee and the ankle. It contains two bones, the tibia and fibula, connected by an interosseous membrane. Use your textbook and skeleton to study and identify the main features of these bones (p. 404). The deep crural fascia of the leg attaches superiorly to the condyles of the tibia and head of the fibula, where it is continuous with the fascia lata of the thigh. Distally, the crural fascia attaches to the medial and lateral malleoli and the calcaneous bone at the ankle. Anterior and posterior intermuscular septa from the crural fascia pass deeply into the leg. These structures, along with the tibia, fibula and interosseous membrane, define the anterior (extensor), lateral (peroneal) and posterior (flexor) compartments. The posterior compartment is further divided into superficial and deep compartments by a transverse septum. At the ankle joint, the crural fascia thickens to form the superior and inferior extensor retinacula anteriorly, the flexor retinaculum medially and the peroneal retinacula laterally. These retinacula will be dissected with the muscles within each compartment.

29. ANTERIOR COMPARTMENT OF THE LEG AND DORSUM OF THE FOOT

Complete the removal of the skin and superficial fascia from the anterior surface of the leg and dorsum of the foot to the toes, as illustrated in Figure 4.3. Identify the structures in the superficial fascia as it is removed. Note the inferior course of the **greater saphenous vein** (p. 444) in the superficial fascia on the medial aspect of the leg. Identify its origin on the dorsum of the foot. The **saphenous cutaneous nerve** is a terminal branch of the femoral nerve, which enters the leg from the distal end of the adductor canal. In the leg, the saphenous nerve parallels the course of the greater saphenous vein. Within the superficial fascia on the lateral lower third of the leg, identify the **superficial peroneal nerve** (pp. 450–451) as it reaches the skin just superior to the ankle. Understand and follow its distribution on the dorsum of the foot. At the ankle, clean and observe the **superior** and **inferior extensor retinacula** (pp. 423; 452), noting their positions and attachments. Review the anterior bony features of the tibia and fibula and identify the tarsal, metatarsal and phalangeal bones of the foot, observing their relationships and articulations.

A. Muscles

The muscles of the anterior compartment of the leg insert on the dorsum and medial side of the foot. Thus, the leg and foot will be studied together. Three of the four muscles that comprise the anterior compartment arise from the tibia, fibula and interosseous membrane. Remove the fascia from the dorsum of the foot and anterior leg to identify these structures. From the medial to the lateral aspect, identify (Figure 4.6) the **tibialis anterior, extensor hallucis longus** and **extensor digitorum longus** muscles (pp. 423; 426; 452–455). The tibialis anterior becomes tendinous at the lower third of the leg. Follow its tendon deep to the two extensor retinacula, where it is covered by a synovial sheath. This tendon reaches the medial side of the foot with its insertion on the medial cuneiform and the base of the first metatarsal bone. The extensor hallucis longus (p. 455) is found on the lateral side of the tibialis anterior in the lower part of the leg. The tendon of the extensor hallucis (p. 455) is covered by a synovial sheath as it passes deep to the extensor retinacula (Figure 4.6). Follow the tendon of the hallucis longus across the foot to its insertion on the base of the distal phalanx of the great toe. The extensor digitorum longus is lateral and passes deep to the inferior retinaculum, which also is covered by a synovial sheath. It divides

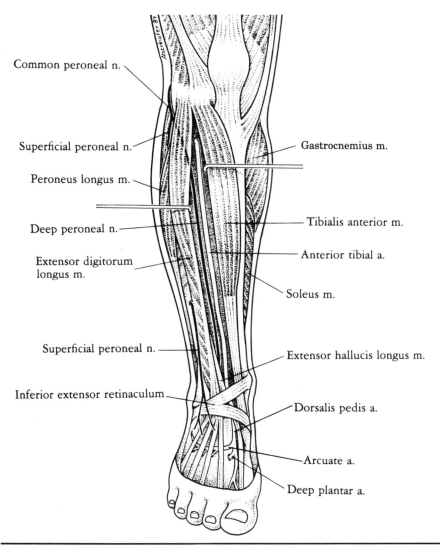

Common peroneal n.

Superficial peroneal n.

Peroneus longus m.

Deep peroneal n.

Extensor digitorum
longus m.

Superficial peroneal n.

Inferior extensor retinaculum

Gastrocnemius m.

Tibialis anterior m.

Anterior tibial a.

Soleus m.

Extensor hallucis longus m.

Dorsalis pedis a.

Arcuate a.

Deep plantar a.

Figure 4.6

into four tendons (pp. 452–455) inferior to the inferior retinaculum. Clean and follow the four tendons across the foot to their terminations on the lateral four toes. On the dorsum of each toe, observe that the extensor tendon divides into two lateral slips and one central slip, which attach to the distal and middle phalanges, respectively. This pattern of insertion forms extensor expansions on the lateral four toes. The fourth muscle of the anterior compartment is the **peroneus tertius** (pp. 452–455). This muscle is actually a lateral slip of the extensor digitorum longus. Follow the peroneus tertius deep to the extensor retinacula and note its insertion on the dorsum of the fifth metatarsal bone. Review the dorsiflexion and inversion actions of the muscles of the anterior compartment. All these muscles are innervated by the deep peroneal nerve, which will be described.

On the dorsum of the foot, note the positions of the **extensor digitorum brevis** and **extensor hallucis brevis** (pp. 452–455), which are intrinsic muscles deep to the tendons of the extensor digitorum longus. The extensor brevis digitorum is thin and passes diagonally across the foot to form four tendons to the medial four toes. The tendon to the great toe is the extensor hallucis brevis. The lateral three tendons of the extensor digitorum brevis participate with the tendons of the extensor digitorum longus in forming the extensor expansions of these digits.

B. Vessels and Nerves

The nerve and vessels of the anterior compartment are the **deep peroneal nerve** and **anterior tibial vessels** (pp. 452–455). They descend the leg on the interosseous membrane between the tibialis anterior and the extensor hallucis longus. Retract these two muscles to identify and follow the nerve and these vessels (Figure 4.6). The deep peroneal nerve innervates all four muscles of the anterior compartment and the extensor digitorum brevis and extensor hallucis brevis in the foot. This nerve is one of the two main divisions of the common peroneal nerve (superficial and deep peroneal nerves). The common peroneal nerve was previously observed following the tendon of the biceps femoris in the popliteal fossa. Continue to follow its course around the neck of the fibula within the upper fibers of the peroneus longus, where it divides into the **deep** and **superficial deep peroneal nerves** (pp. 452–455). Cut open the superior fibers of the peroneus longus to observe this division. The deep peroneal nerve descends on the lateral side of the tibialis anterior on the interosseous membrane in the anterior compartment, innervating the muscles of this compartment. In the foot, the deep peroneal nerve is cutaneous to the dorsum of the great toe and adjacent second toe and motor to the intrinsic extensor digitorum brevis and extensor hallucis brevis muscles on the dorsum of the foot, as discussed previously. The superficial peroneal nerve will be dissected in the lateral compartment of the leg.

The **anterior tibial artery** branches from the popliteal artery at the inferior border of the popliteus muscle in the lower aspect of the popliteal fossa. This origin in the posterior leg will be dissected later. The anterior tibial artery then enters the anterior compartment by passing superior to the interosseous membrane. In the anterior compartment, follow the anterior tibial artery as it descends medial to the deep peroneal nerve. This artery supplies muscles of the leg and provides anastomotic branches at the knee and ankle joints. As the artery passes deep to the inferior extensor retinaculum, it enters the dorsum of the foot, where it is called the **dorsalis pedis artery** (p. 455). Follow this artery across the foot and identify distally the **arcuate, deep plantar** and **dorsal metatarsal arteries.** These arteries are mostly deep to the extensor digitorum brevis. Other branches are anastomotic at the ankle joint (lateral and medial malleolar arteries, lateral and medial tarsal arteries).

30. LATERAL COMPARTMENT OF THE LEG

The lateral aspect of the leg contains two muscles and the superficial peroneal nerve (pp. 423; 452–455). Identify the **peroneus longus** and **peroneus brevis** muscles on the lateral side of the leg (Figure 4.6). Separate and follow their tendons inferiorly (the peroneus longus is most posterior) where they pass in a groove posterior to the lateral malleolus. The tendons of these muscles are covered by synovial sheaths and are bounded by the **superior** and **inferior peroneal retinacula** at the malleolus. Identify these retinacula, noting their attachments. Distal to the retinacula, the peroneus longus turns medially around a groove in the cuboid bone, courses medially in a deep plane across the sole of the foot and finally attaches to the lateral surface of the first metatarsal and medial cuneiform bones. This deep course of the tendon of the peroneus longus cannot be observed until the dissection of the sole of the foot. The peroneus brevis remains on the lateral side of the foot and passes to the tuberosity at the base of the fifth metatarsal bone. These two muscles evert the foot and assist in plantar flexion.

The **superficial peroneal nerve** branches from the **common peroneal nerve** at the neck of the fibula (p. 452). Identify the superficial peroneal nerve as it passes inferiorly between the two peroneal muscles, which it innervates. Distally, it enters the dorsal aspect of the foot as a cutaneous nerve (p. 451).

31. POSTERIOR COMPARTMENT OF THE LEG

Continue to reflect the skin on the posterior surface of the leg and ankle and onto the plantar surface of the foot, as illustrated in Figure 4.1. Carefully dissect away the superficial fascia of the leg and identify the **lesser saphenous vein** and **medial sural cutaneous nerves** (p. 447). The lesser saphenous vein passes from the dorsum of the foot inferior to the lateral malleolus and ascends the middle of the dorsal surface of the leg to terminate in the popliteal vein within the popliteal fossa. Identify and clean the borders of the **flexor retinaculum** at the medial malleolus of the ankle (pp. 422;

425). The posterior compartment is divided into superficial and deep compartments.

A. Superficial Compartment

The muscles of the superficial compartment (Figure 4.7) include the gastrocnemius, soleus and plantaris (pp. 421–422; 447–449). These muscles insert via the calcaneal tendon on the tuberosity of the calcaneus and are powerful plantar flexors of the foot. They are innervated by the tibial nerve. The **gastrocnemius** is the most superficial muscle of this group, forming most of the bulk and contour of the calf.

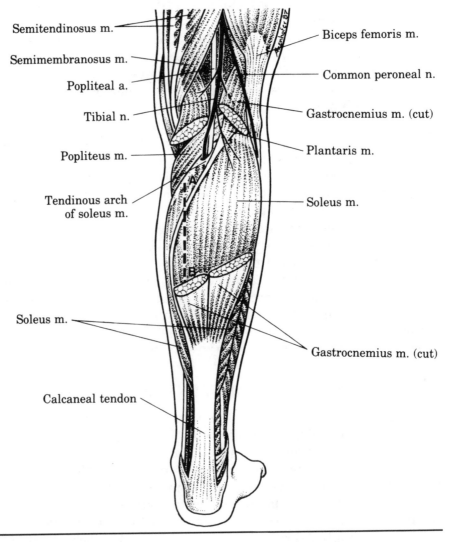

Semitendinosus m.

Semimembranosus m.

Popliteal a.

Tibial n.

Popliteus m.

Tendinous arch of soleus m.

Biceps femoris m.

Common peroneal n.

Gastrocnemius m. (cut)

Plantaris m.

Soleus m.

Soleus m.

Gastrocnemius m. (cut)

Calcaneal tendon

Figure 4.7

Review its lateral and medial heads of origin and their relationships to the popliteal fossa. Note the courses of the tibial nerve and popliteal vessels deep to the two heads of the gastrocnemius from the popliteal fossa. Clean and follow the gastrocnemius inferiorly to the calcaneal tendon and not its insertion on the calcaneus. Note the **plantaris** as it courses with the lateral head of the gastrocnemius (p. 421). It has a small, fleshy portion that can be identified superior to the origin of the lateral head of the gastrocnemius. Its long tendon passes deep to the gastrocnemius to reach the calcaneal tendon. The **soleus** lies deep to the gastrocnemius and can be observed better after cutting the two heads of the gastrocnemius just before they fuse in the midline. Reflect the gastrocnemius inferiorly and note the broad, fleshy fibers of the soleus. The soleus is a thin muscle that has a wide origin from the fibula and soleal line of the tibia. A fibrosus arch extends between these two bones. The tibial nerve and popliteal vessels leave the popliteal fossa by passing deep to the fibrous arch of the soleus (p. 446) to enter the posterior compartment. Clean the borders and follow the soleus inferiorly to the point where it joins the calcaneal tendon. Note the tendon of the plantaris.

B. Deep Compartment

The structures within the deep compartment are better observed after reflecting the tibial (medial) attachment of the soleus (Figure 4.7, A-B). Clean and identify the tibial fibers of the soleus that attach to the soleal line. Carefully cut these fibers vertically from the tibia and reflect them laterally to expose the deeper area of the posterior leg (p. 449). The line of origin of the soleus from the tibia can be extensive.

The muscles of the deep compartment (Figure 4.8) include the **flexor hallucis longus, flexor digitorum longus, tibialis posterior** and **popliteus** (pp. 424–425; 448–449). The first three muscles plantar flex the foot and the popliteus is a weak flexor of the knee joint. The flexor hallucis longus and flexor digitorum longus also flex the toes. All these muscles are innervated by the tibial nerve.

Identify the **popliteus** muscle in the floor of the popliteal fossa (p. 421) deep to the popliteal artery. Note that the popliteus is a thin, triangular muscle that arises from the lateral condyle of the femur. The tendon of the popliteus passes within the capsule of the knee joint between the lateral meniscus and the capsule. Next, identify the **flexor hallucis longus** on the lateral side of the leg at the fibula, the **flexor digitorum longus** on the medial of the leg at the tibia and the **tibialis posterior** lying deeply between the flexor hallucis longus and the digitorum longus. Follow the tendons of these three muscles posterior to the medial malleolus as they enter the foot deep

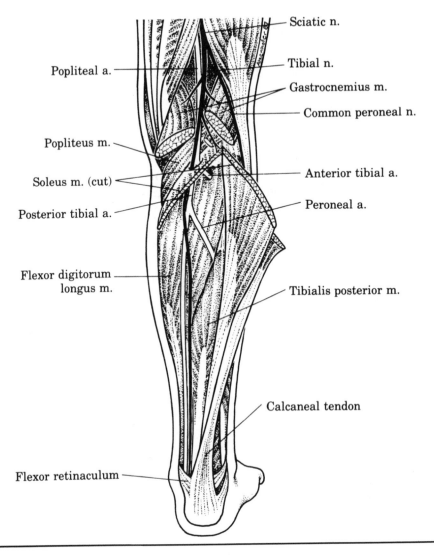

Popliteal a.

Popliteus m.

Soleus m. (cut)

Posterior tibial a.

Flexor digitorum
longus m.

Flexor retinaculum

Sciatic n.

Tibial n.

Gastrocnemius m.

Common peroneal n.

Anterior tibial a.

Peroneal a.

Tibialis posterior m.

Calcaneal tendon

Figure 4.8

to the **flexor retinaculum,** which extends between the medial malleolus and the calcaneus. Reflect the skin from the lateral and medial sides of the ankle. These three muscles develop synovial sheaths as they pass deep to the retinaculum. Cut the retinaculum and observe that posterior to the medial malleolus, the tibialis posterior is the anterior muscle, the flexor digitorum longus is in the middle, and the flexor hallucis longus is posterior. The posterior tibial vessels and nerve are between the flexor digitorum longus and the flexor hallucis longus. These three tendons then enter the sole, where their courses and insertions will be studied later. Clean all the fascia on the medial side of the ankle at the flexor retinaculum to observe these relationships.

The **tibial nerve** (pp. 445–446) leaves the popliteal fossa and enters the posterior compartment of the leg deep to the gastrocnemius and fibrous arch of the soleus. Follow the course of the tibial nerve as it descends vertically on the tibialis posterior through the center of the leg (Figure 4.8). During its course, the nerve crosses the posterior tibial artery from its medial to its lateral side. The tibial nerve provides branches to the muscles of the deep compartment as well as cutaneous branches to the heel and articular branches to the ankle joint. Follow the tibial nerve deep to the flexor retinaculum, where it lies between the posterior tibial artery and vein anteriorly and the flexor hallucis longus posteriorly.

Next, clean and identify the **posterior tibial artery** and **vein** (p. 449) as they enter the leg from the popliteal fossa with the tibial nerve. The popliteal artery passes deep to the fibrous arch of the soleus. At the inferior border of the popliteus, identify the division of the popliteal artery into the **anterior** and **posterior tibial arteries.** The anterior artery was observed in the anterior compartment (Figure 4.8). The posterior tibial artery is the continuation of the popliteal artery distal to the origin of the anterior tibial artery. The posterior tibial artery courses diagonally through the leg to pass posterior to the medial malleolus deep to the flexor retinaculum. In the leg, the posterior tibial artery provides many muscular branches and several collateral branches at the ankle. Identify the **peroneal artery,** the largest branch of the posterior tibial artery in the leg. Note that it arises 2 cm inferior to the origin of the anterior tibial artery and passes to the lateral (fibula) side of the leg. It then descends on or in the flexor hallucis to the lateral side of the ankle. The peroneal artery provides many of the main muscular arteries to the posterior and lateral compartments. At the ankle, it provides perforating and communicating branches to the anterior and posterior tibial arteries and a posterior lateral malleolar branch, all of which participate in collateral circulation at the ankle. The **posterior tibial** and **peroneal veins** course with their corresponding arteries and should be dissected with them. Their tributaries are similar to those of the arteries.

32. PLANTAR SURFACE OF THE FOOT

Review the bony structures of the foot before beginning the dissection of the soft structures (pp. 406–407). Complete the removal of the skin from the sole distally to the nail bed, as illustrated in Figure 4.1, noting the thickness of the skin as it is removed. Deep to the skin of the sole, the deep fascia is thickened considerably to form the **plantar aponeurosis** (pp. 427; 456). Remove this aponeurotic layer by cutting the septa that penetrate and attach deeply to the bones of the foot. Note the increased thickness of its central part, which is tightly anchored to the calcaneal tuberosity posteriorly and bases of the toes anteriorly. These attachments help support the longitudinal arch of the foot. Cutaneous nerves to the skin penetrate the aponeurosis. The plantar aponeurosis with the deep muscular septa divide the plantar surface into four compartments similar to those of the hand. The compartments are the great toe, small toe, central and interosseous-adductor. Appreciate this organization during the dissection. After removal of the aponeurosis, the four layers of the foot can be dissected.

A. First Layer

This layer (Figure 4.9) is immediately deep to the plantar aponeurosis and is composed of three muscles (p. 427). Identify the **abductor digiti minimi** in the small-toe compartment on the lateral aspect of the foot. It arises from the calcaneus and courses distally to insert on the base of the proximal phalanx of the fifth toe. Medially in the great-toe compartment, identify the **abductor hallucis,** which arises from the calcaneus and inserts on the base of the proximal phalanx of the first (great) toe. In the first layer within the central compartment is the **flexor digitorum brevis,** which lies deep to the plantar aponeurosis and arises from its and the calcaneus. Note that the flexor digitorum brevis divides into four tendons that insert on the lateral four toes by separating into two lateral slips that attach to the middle phalanges. Follow these tendons into the toes and note that the tendons of the flexor digitorum longus insert on the distal phalanx by passing between the lateral slips of the flexor digitorum brevis (pp. 427; 456). To follow these tendons into the toes, open the **digital fibrous** (*tendon*) **sheaths** that cover the tendons of the flexor digitorum brevis and longus. The tendons are covered by synovial sheaths. Review the actions and attachments of the muscles of this layer.

Deep to the first layer are the initial courses of the main nerves and vessels to the foot. These nerves and vessels can be observed after cutting and reflecting the flexor digitorum brevis at its midpoint (p. 457). Identify and separate completely the vessels, nerve and tendons that pass posterior to the medial malleolus and deep to the flexor retinaculum to enter the foot by cutting the flexor retinaculum. Deep to the retinaculum, the vessels (posterior tibial) and nerve (tibial) enter the fibers of the abductor hallucis. Cut and reflect

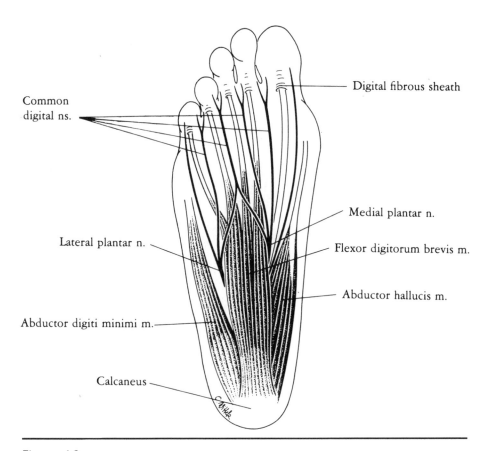

Common
digital ns.

Digital fibrous sheath

Medial plantar n.

Lateral plantar n.

Flexor digitorum brevis m.

Abductor hallucis m.

Abductor digiti minimi m.

Calcaneus

Figure 4.9

the fibers of the abductor hallucis to establish continuity between the posterior leg and the sole. Carefully tease through the fibers of origin of the abductor hallucis to complete the courses of these structures to the foot. From the **tibial nerve** and **posterior tibial vessels,** note the formation of the **medial** and **lateral plantar nerves** and **vessels** within the fibers of the abductor hallucis.

Follow the **medial plantar nerve** (p. 457) from its origin as it courses distally in the foot between the abductor hallucis and the flexor digitorum brevis (Figure 4.9). It innervates these two muscles and the deeper flexor hallucis brevis and first lumbrical muscles. Cutaneous branches (proper and common digital branches) are provided to the medial three and one-half toes. Identify the **medial plantar artery** as it courses with the nerve. This artery is extremely small and arises from the posterior tibial artery deep to the abductor hallucis. The medial plantar artery then passes distally to supply the medial muscles and digital branches to the great toe.

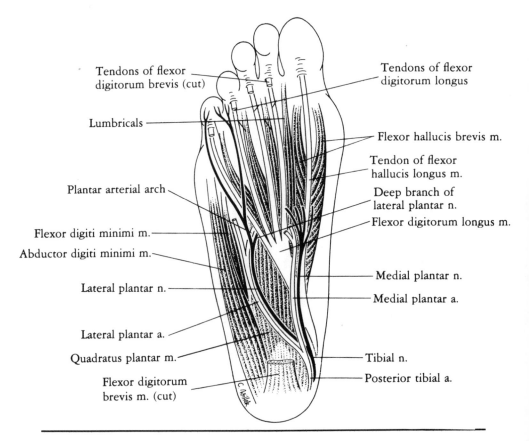

Figure 4.10

Identify the **lateral plantar nerve** (pp. 456–457) as it crosses to the lateral side of the sole in the fascial space between the **flexor digitorum brevis** superficially and the **quadratus plantae** deeply (Figures 4.9 and 4.10). This nerve is medial to its accompanying artery. The lateral plantar nerve first innervates the abductor digiti minimi and quadratus plantae. Note that lateral to the quadratus plantae, it divides into a superficial branch and a deep branch. The **superficial branch** supplies cutaneous digital branches to the lateral one and one-half toes and motor fibers to the flexor digiti minimi brevis. The **deep branch** enters the interosseus-adductor layer of the foot and will be dissected with that area. Next, identify the **lateral plantar artery,** a branch of the posterior tibial artery that courses posterior (lateral) to the nerve. It supplies blood to the lateral aspect of the foot. The lateral plantar artery divides distally into a **superficial branch** to the lateral toes and a **deep branch** that forms the **plantar arterial arch,** which will be dissected in the third layer.

B. Second Layer

The second layer (Figure 4.10) of muscles includes the quadratus plantae, lumbrical muscles and the tendon of the flexor digitorum longus (pp. 428; 457). Identify the **quadratus plantae** deep to the lateral plantar nerve and vessels. It attaches to the calcaneus posteriorly and to the oblique position of the tendon of the flexor digitorum longus anteriorly. It is innervated by the lateral plantar nerve. Review the relationships of the flexor digitorum longus as it passes posterior to the medial malleolus to enter the plantar surface. The tendon of the **flexor digitorum longus** then passes inferior to the tendon of the flexor hallucis longus and crosses the foot obliquely, where it receives the insertions of the quadratus plantae (p. 425). Follow distally the tendons of the flexor hallucis longus and flexor digitorum longus into their respective digits, noting their patterns of insertion. There are usually four small **lumbrical** (p. 428) muscles that arise from the tendon of the flexor digitorum longus (p. 428). The number of lumbrical muscles is highly variable. Identify the origins of the lumbrical muscles and follow them into the toes on the plantar surface of the deep transverse metatarsal ligament, where they insert into the dorsal extensor expansion of the lateral four toes. The most medial lumbrical muscle is innervated by the medial plantar nerve and the lateral three lumbrical muscles are supplied by the deep branch of the lateral plantar nerve. Review the innervations, actions and attachments of this muscle layer.

C. Third Layer

To identify this layer (Figure 4.11), cut and reflect the tendon of the flexor digitorum longus (with the attached quadratus plantae) proximal to its division into four tendons (pp. 429; 458). Deep to this reflection (Figure 4.11), identify the **flexor hallucis brevis** muscle, which divides into two heads that each contain a sesamoid bone and surround the tendon of the flexor hallucis longus; the **flexor digiti minimi** deep to the abductor digiti minimi; and the **adductor hallucis.** The latter muscle has **transverse** and **oblique heads,** which should be cleaned and identified. Reflect the oblique head to expose the deeper course of the **plantar arterial arch** and the **deep branch of the lateral plantar nerve** (p. 458) (Figure 4.11). This artery and nerve lie on the deep interosseous muscles. The deep branch of the lateral plantar nerve innervates the lateral three lumbrical muscles, interosseous muscles and adductor hallucis. Identify the plantar arterial arch (Figure 4.11) and note that it is formed laterally by the deep branch of the lateral plantar artery, which joins the deep plan-

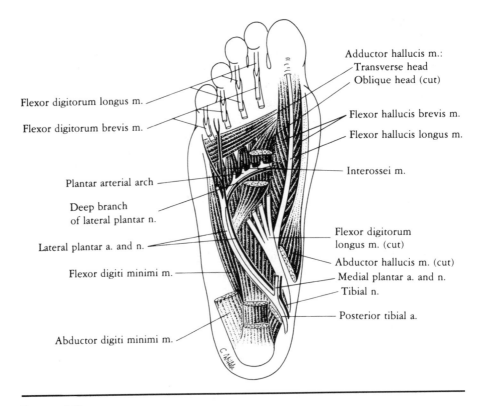

Figure 4.11

tar branch of the dorsalis pedis artery medially. The deep plantar artery can be identified as it penetrates the first intermetatarsal space on the dorsum of the foot. As the plantar arch passes across the interosseous muscles, it gives rise to four **plantar metatarsal arteries.** These arteries pass distally into the toes to form the plantar digital arteries.

D. Fourth Layer

The deepest muscle layer is composed of **four dorsal** and **three plantar interosseous** muscles (Figure 4.11). These muscles function around the axis of the second digit. Identify these muscles as dorsal or plantar by observing the side of the digit on which they insert. Review the innervations and actions of these and the lumbrical muscles on the digits (pp. 429; 458). The function of the interosseous muscles in the foot is much reduced from the action of those of the hand.

33. LIGAMENTS OF THE PLANTAR SURFACE OF THE FOOT

The stability and positions of the bones of the foot are maintained by a number of ligaments that span the bones. These ligaments are deep to the fourth layer of muscles and are closely related to the oblique course of the tendon of the peroneus longus. Dissection of these ligaments is usually possible only if the deepest muscles are removed (pp. 413–414). Identify the ligaments described in the text that follows on only one foot.

A. Plantar Calcaneonavicular (Spring) Ligament

This ligament spans the joint between the talus and the navicular bones and is firmly attached to the sustentaculum tali. Along with the tibialis anterior and posterior tendons, this ligament helps support the longitudinal arch of the foot.

B. Long Plantar Ligament

This ligament spans the central and lateral aspects of the foot between the calcaneous posteriorly and the cuboid bone and bases of the lateral four metatarsal bones anteriorly. As it passes distally, the long plantar ligament covers and forms a tunnel for the tendon of the peroneus longus. Both structures are also important for support of the longitudinal arch. Identify the **tendon of the peroneus longus** deep to this ligament.

C. Plantar Calcaneocuboid Ligament

This ligament is deep to the long plantar ligament and extends anteriorly to the cuboid bone. It is broad and strong and supports the longitudinal arch.

34. ARTICULATIONS

Descriptions of the major joints of the lower limb are provided for courses that allow time for these dissections. This work should be done on only one limb. You should understand and review the articulating surfaces of the bones of each joint before beginning these dissections.

A. Hip Joint

This synovial ball-and-socket joint is designed primarily for strength and stability but also permits a significant degree of movement. Study the bony aspects of the upper end of the femur and acetabulum of the os coxae, which comprise the articulating surfaces (p. 402). To approach the joint anteriorly, cut and reflect the sartorius, rectus femoris, pectineus and iliopsoas muscles. Examine the anterior surface of the capsule. Review its attachments to the margin of the acetabulum superiorly and to the intertrochanteric line and crest inferiorly. The capsule is extremely strong and its fibers course in a spiral fashion, becoming twisted during extension (pp. 408–409). Anteriorly, identify the thickening of the capsule to form the **iliofemoral ligament** and note its function during movement of the lower limb. Next, open the joint cavity by cutting the anterior capsule and examine the internal articulating surfaces. Observe the **ligament of the head of the femur.** The posterior aspects of the joint can be observed by reflecting the obturator internus, gamelli, piriformis, quadratus femoris and gluteus medius and minimus in the gluteal region. This maneuver provides a view of the **ischiofemoral ligament.** Study the extent of the synovial membrane and the blood supply of the cavity. Review the movements and functions of each muscle that acts upon the hip joint.

B. Knee Joint

Articulation at the knee joint provides important weight-bearing functions, with freedom of flexion and extension and slight rotation. Stability is provided by large articulating surfaces (condyles) of the tibia and femur, a strong capsule and strong collateral and intra-articular ligaments. Describe the extent of the capsule and synovial membrane. Review the articulating surfaces of the condyles of the tibia and femur (pp. 410–411) before beginning the dissection.

On the lateral side of the knee, follow the **iliotibial tract** to its attachment on the lateral condyle of the tibia. Cut the attachment of the adjacent tendon of the biceps femoris to the head of the fibula and identify the cordlike **fibular collateral ligament.** This ligament extends between the lateral condyle of the femur and the head of the fibula and prevents lateral displacement of the leg. Note that the popliteus separates this ligament from the synovial membrane and lateral meniscus.

On the medial side of the knee, retract and cut the sartorius, gracilis and semitendinosus at their insertions. Deep to these muscles, identify the **tibial collateral ligament** (p. 410), which strengthens the joint medially. The tibial collateral ligament extends between the medial condyles of the femur and tibia and is firmly attached to the medial meniscus deeply. Both the tibial and the fibular collateral ligaments attach posterior to the vertical axis of the joint and thus become tight during extension.

Anteriorly, review the insertions of the quadriceps muscles to the patella and then to the tibia via the **patellar ligament.** Cut the insertions of these muscles to the patella and then make a transverse cut through the deeper anterior capsule to open the joint cavity (p. 410). Flex the knee and observe the large **synovial cavity, anterior** and **posterior cruciate ligaments** and **medial** and **lateral menisci.** Note the positions of the two ligaments and menisci and review the functions of these internal structures. Posterior to the knee, examine the relationships of the **popliteus** muscle and tendon and **posterior meniscofemoral ligament.** Review the movements of the joint and actions of the muscle at the knee.

C. Ankle Joint

This hinge joint is between the distal ends of the tibia (medial malleolus) and fibula (lateral malleolus) and the convex surface of the **trochlea of the talus** (p. 413). The capsule is weak anteriorly and posteriorly but thickens on each side to form collateral ligaments on its medial and lateral sides (pp. 413–414). Medially, identify the triangular **deltoid ligament,** which extends between the medial malleolus and the talus, navicular bone and calcaneus. Laterally, ligaments can be observed attaching the lateral malleolus to the talus and calcaneus. These lateral ligaments are weaker than the deltoid ligament. Review the actions of dorsiflexion and plantar-flexion at the joint.

D. Joints of the Foot

The important movements of inversion and eversion occur primarily at the midtarsal, talonavicular and calcaneocuboid joints (pp. 413–414). Examine these bones on an articulated foot and identify the subtalar joint at the lower aspect of the talus. Review the relationships of the ligaments identified earlier on the deepest aspect of the foot to these articulations. Describe the supports for the transverse and longitudinal arches of the foot.

CHAPTER FIVE

THORAX

35. SURFACE ANATOMY

Observe and palpate the major surface landmarks on the anterior and lateral chest walls before beginning the dissection of the thorax. In the superior midline, note the **suprasternal notch** (jugular notch) at the superior surface of the **manubrium sterni** with the sternal articulations of the clavicles and attachments of the sternocleidomastoid muscles on both sides. Approximately 5 cm inferior to the suprasternal notch is a transverse ridge where the manubrium sterni articulates with the **body of the sternum.** This ridge is the **sternal angle (of Louis),** one of the most reliable landmarks on the anterior chest wall. The sternal angle demarcates the position of the articulation of the second costal cartilage with the sternum. Inferiorly from the sternal angle, the remaining ribs can be counted. However, the eleventh and twelfth ribs are difficult to identify and the first ribs, being deep to the clavicles, cannot be easily palpated. At the inferior end of the sternum are the cartilaginous **xiphoid process** and **xiphisternal junction.** This junction indicates the anterior surface projection of the inferior border of the heart, upper surface of the liver and anterior midline attachment of the diaphragm. The nipples are usually located in the fourth intercostal space. However, they are not always reliable landmarks because of the variable amount of fatty tissue and the highly variable size of the breasts in the female. At this time, make a complete study of the bony elements that comprise the thoracic wall (pp. 174–181).

36. THORACIC WALL

Removal of the mammary gland and dissection of the pectoral muscles and fasciae were completed during the dissection of the upper limb (see Chapter 3, Section 12). If this work has not been completed, it should be finished before the dissection of the anterior chest wall is begun. Review the anterior and lateral cutaneous nerves previously observed when the skin was reflected from the chest wall.

To begin the dissection of the intercostal muscles, detach the more-anterior fibers of the **serratus anterior** muscle from the upper eight ribs to expose the intercostal spaces. In the anterior aspect of one or two representative intercostal spaces on each side of the body, clean and identify the **external intercostal** muscles, which form the most superficial muscle layer in the intercostal spaces (Figure 5.1). There are 11 pairs of these muscles, which extend from the tubercles of the ribs posteriorly to the costochondral articulations anteriorly. From this anterior limit, a thin external intercostal membrane extends to the sternal margin and completes the superficial layer. Note the

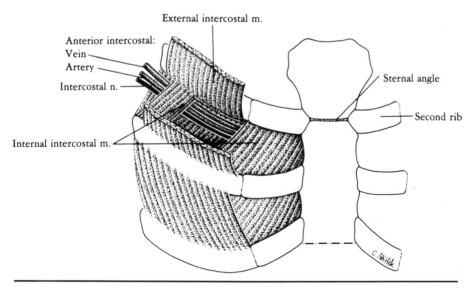

External intercostal m.

Anterior intercostal:
Vein
Artery
Intercostal n.

Sternal angle

Second rib

Internal intercostal m.

Figure 5.1

attachments and directions of fibers of the external muscles and membrane (pp. 185; 189).

Divide and remove the external intercostal muscles from several intercostal spaces to expose the deep **internal intercostal** muscles (pp. 185; 189). There are also 11 pairs of these muscles, which extend from the lateral margin of the sternum anteriorly to the angle of the ribs posteriorly. Note the attachments and directions of these muscle fibers and contrast them with the external intercostal muscles (Figure 5.1). From the angle of the ribs posteriorly, there is an internal intercostal membrane that continues this layer medially to the tubercle of the rib. These posterior relationships are not in view at this time but will be seen during dissection of the posterior thoracic wall.

Dissection of the intercostal nerves, arteries and veins can be accomplished best after removal of the anterior portion of the chest wall (Figure 5.2). Make a shallow, transverse incision through the muscles of the sixth intercostal space (do not cut deeply into the pleura) from the midaxillary line on each side toward the midline. As this cut is being made, identify the pleural membrane on the inside of the chest wall. Use your hand to push the pleura and lungs away from the chest wall (pp. 238–239) to avoid disturbing them deep to the ribs. Next, with bone forceps, cut through the sixth to second ribs slightly posterior to the midaxillary line on each side. Cuts must be made slightly more anterior for the second and third ribs to leave the contents of the axilla undisturbed. Continue the initial transverse

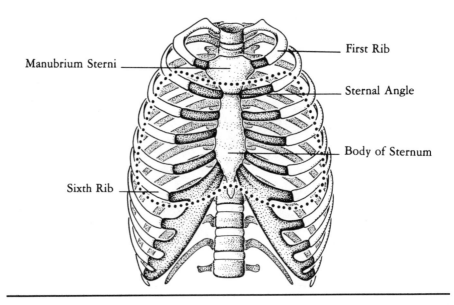

Figure 5.2

cut in the sixth intercostal space across the midline at the xiphister-nal junction. The **internal thoracic arteries** and **veins** course on the deep aspect of the lateral margins of the sternum (pp. 189; 239). Better relationships can be maintained if one pair of these vessels can be removed with the chest wall on one side and retained in the cadaver on the other side. To dissect in this manner, gently elevate the chest wall from below at the sixth intercostal space in the mid-line. Use your hand and fingers to develop the space between the pleura and the endothoracic fascia. This fascia lines and is fused to the inner aspect of the rib cage (p. 189). Elevating the chest wall at the level of the xiphisternal junction exposes the **transversus thoracis** muscle, which holds the internal thoracic vessels against the costal cartilages. Break through this muscle on the side where the vessels are to be left intact in the cadaver. On the opposite side, cut through the vessels at the level of the xiphisternal junction and remove them with the chest wall. As the chest wall is being pulled superiorly, continue to separate the pleura from the chest wall by using your fingers to keep the pleura intact as much as possible. After elevating the anterior chest wall superiorly to the second inter-costal space, cut through the muscle of this space on each side toward the manubrium sterni. With bone cutters, make a transverse cut across the midpoint of the sternum between the suprasternal notch and the sternal angle (Figure 5.2). Again, be careful to identify and avoid cutting the internal thoracic vessels at the superior end of the

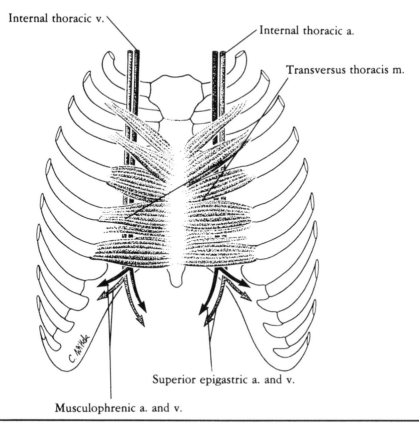

Internal thoracic v.

Internal thoracic a.

Transversus thoracis m.

Superior epigastric a. and v.

Musculophrenic a. and v.

Figure 5.3

sternum on the side where they will be left in the cadaver. The entire anterior chest wall can now be removed.

The two **internal thoracic arteries** (Figure 5.3) can now be seen on the inner aspect of the anterior thoracic wall as they descend between the upper five costal cartilages and the pleura (pp. 190; 238–239). These arteries originate from the right and left subclavian arteries in the root of the neck and enter the thorax just posterior to the sternoclavicular joints. At the inferior end of the sternum, the internal thoracic artery is separated from the pleura by the transversus thoracis muscles and divides posterior to the sixth costal cartilage into the **superior epigastric** and **musculophrenic arteries** (Figure 5.3). The internal thoracic artery is accompanied by two **internal thoracic veins,** which drain into the brachiocephalic veins. Note that on the internal surface of the chest wall, the **transversus thoracis** arises from the inner surface of the sternum (Figure 5.3). The fibers of this muscle pass upward and laterally and attach to the inner surfaces of the second to sixth costal cartilages.

Carefully dissect into the inner aspects of several costal grooves on the inferior aspect of the ribs in the intercostal spaces. In the grooves, identify the **anterior intercostal vessels** and **intercostal nerves** (pp. 189; 196). The anterior arteries and veins are small and often difficult to locate. The anterior intercostal arteries (usually two in each space) branch from the internal thoracic artery in the upper five spaces or from the musculophrenic arteries in the lower six spaces. The anterior intercostal arteries continue to course laterally through the intercostal spaces, where they anatomose posteriorly with the posterior intercostal (and collateral intercostal) arteries. The posterior vessels will be described with the posterior thoracic wall in the posterior mediastinum. The intercostal veins (anterior and posterior) accompany the arteries and connect the internal thoracic and musculophrenic veins anteriorly to the azygos system of veins posteriorly. The veins are the most superior structures in the costal grooves. The **intercostal nerves** (T1 to T11 ventral rami) are easier to identify. They provide somatic and motor innervation to the muscles and skin of the chest wall and to the abdominal wall for nerves 7 to 11. Review the distributions of the lateral cutaneous branches in the midaxillary line and of the anterior cutaneous branches at the sides of the sternum. In the costal groove (Figure 5.1), the artery lies between the vein superiorly and the nerve inferiorly (p. 254). The vessels and nerves will be more easily identified later on the posterior aspects of the intercostal spaces.

37. THORACIC CAVITY AND PLEURA

With the anterior chest wall removed, first explore the viscera of the thorax *in situ* (pp. 238–243). Note the relationship between the laterally placed lungs and the pleurae to the central mediastinum (Figure 5.4). The mediastinum contains most of the thoracic viscera and is limited by the sternum anteriorly, the 12 thoracic vertebral bodies posteriorly and the parietal pleura laterally. Understand the division of the mediastinum into superior, anterior, middle and posterior compartments. The thoracic cavity is continuous superiorly with the neck through the **superior thoracic aperture** and is separated inferiorly from the abdomen by the diaphragm, which closes the **inferior thoracic aperture.**

The **pleura** surrounds each lung and is composed of two layers (Figure 5.4), which are continuous with each other at the root of the lung. The inner layer, the **visceral pleura,** is directly adherent to the surfaces of the lung. The outer layer, the **parietal pleura,** is apposed to and follows the contour of the thoracic wall (pp. 238–239).

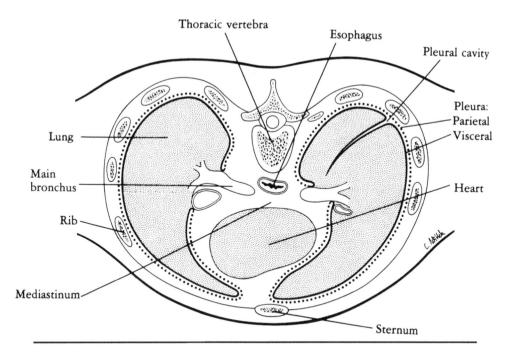

Figure 5.4

The potential space between these two layers, the pleural cavity, contains a small amount of serous fluid, which facilitates movement of the layers against each other during respiration. With your hand and blunt dissection, completely free and separate the parietal pleura from the pericardium, which contains the heart. Follow the courses of the **phrenic nerve** and **pericardiacophrenic vessels** between the pleura and the pericardium. The regions of the parietal pleura are designated as **mediastinal, diaphragmatic, cervical** and **costal** and are named according to the area to which they are related. Note these relationships. The mediastinal and costal parietal pleurae continue over the apices of the lungs to form the cupula of the pleura. Review in your textbook the lines of projection of the parietal and visceral pleurae to the anterior, lateral and posterior chest walls and follow these surface projections as closely as possible on the cadaver (p. 224). Next, make a vertical incision in the parietal pleura on each side of the thorax in the midclavicular line. Place your hand in the pleural cavity and gently explore its limits. Note the superior limit of the cervical pleura at the neck of the first rib and identify the **costomediastinal** and **costodiaphragmatic recesses.** These recesses are important because they provide areas where excess pleural fluid can be withdrawn without damaging the lungs.

38. SUPERIOR MEDIASTINUM

Before proceeding with the dissection, review the divisions of the mediastinum and define the limits and boundaries of each division. The superior mediastinum is the upper portion of the mediastinal space and communicates through the superior thoracic aperture with the neck. It contains structures that course between the thorax and the neck and between the thorax and the upper limb.

To dissect the superior mediastinum, make a midline *vertical* cut through the remaining superior half of the manubrium sterni. Better exposure can be achieved by cutting through the first ribs about 1 to 2 cm lateral to the sternum, being careful not to damage the subclavian vessels where they cross the first rib. With force, pry the two halves of the manubrium upward and laterally.

The most anterior structure in the superior mediastinum is the **thymus gland** (pp. 239–240). In the adult, most of the glandular elements of its two lobes have been replaced by fat and areolar tissue. The lobes lie directly posterior to the manubrium sterni and anterior to the brachiocephalic veins and superior vena cava. Thymic veins are often seen terminating in the left brachiocephalic vein. Bluntly remove the gland and the adjacent fascia and connective tissue that cover the superior mediastinum.

The dissection of the superior mediastinum (Figure 5.5) primarily involves removing the loose connective tissue and separating the structures from each other. *Do not* carry your dissection superiorly into the root of the neck. Note that the superior extent (apex) of the pericardium terminates at the level of the sternal angle, where it becomes continuous with the adventitial layer of the great vessels (ascending aorta, superior vena cava and pulmonary trunk and veins).

After removing the thymic and other loose connective tissue, identify the **right** and **left brachiocephalic veins** (pp. 240–243). The brachiocephalic veins (Figure 5.5) begin posterior to the sternal end of the clavicles by the junction of the internal jugular and subclavian veins in the root of the neck. Expose the brachiocephalic veins and watch for the **internal thoracic, inferior thyroid, vertebral** and **first posterior intercostal veins,** which terminate into each of the veins (p. 243). In addition, the **left superior intercostal vein** is a tributary to the left brachiocephalic vein. These tributaries are variable and all of them may not be identified. The right brachiocephalic vein descends vertically along and deep to the right border of the sternum and the costal cartilages, whereas the left vein passes obliquely and inferiorly across the superior mediastinum to join the right brachiocephalic vein deep to the right first costal cartilage.

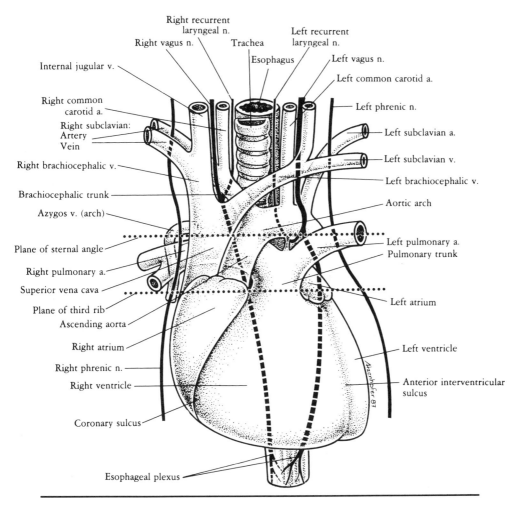

Figure 5.5

From this junction, the **superior vena cava** is formed (pp. 243–245). Identify the superior vena cava as it descends and enters the pericardium at the level of the sternal angle, where it continues its course in the middle mediastinum to terminate in the right atrium. Observe the termination of the **arch of the azygos vein** into the right aspect of the superior vena cava just before it pierces the pericardium (Figure 5.5).

Immediately deep to the brachiocephalic veins are the **aortic arch** and its branches (Figure 5.5) (pp. 245–246; 248–249). It may be useful to reflect the left brachiocephalic vein at its midpoint to gain a better view of these arteries. The aortic arch begins as the continuation of the ascending aorta at the sternal angle. It arches superiorly, posteriorly and to the left deep to the lower half of the manubrium sterni in the superior mediastinum. The posterior course brings the arch into direct contact with the anterior and left sides of the trachea and with the left side of the body of the fourth thoracic vertebra, where it then enters the posterior mediastinum as the descending aorta. Identify the origins and courses of the three branches of the aortic arch. The first and largest branch of the aortic arch is the **brachiocephalic trunk,** which arises posterior to the middle of the manubrium sterni. Follow this trunk to the level of the right sternoclavicular joint. To the left and posterior to the brachiocephalic trunk, the **left common carotid artery** branches from the highest aspect of the aortic arch. Distal to the left common carotid artery, the **left subclavian artery** arises from the arch at the left side of the trachea. Follow the courses of these two arteries to the left, noting their relationships. Do not dissect these three arteries above the level of the clavicle to avoid disturbing the root of the neck. Identify the **ligamentum arteriosum** on the inferior surface of the aortic arch opposite the origin of the left subclavian artery. This band of connective tissue is the remnant of the fetal ductus arteriosus and attaches to the left pulmonary artery in the middle mediastinum (Figure 5.5).

After all these vessels have been identified, follow the courses of the **phrenic** and **vagus nerves** (Figure 5.5), noting their relationships to the great vessels and trachea (pp. 245–246; 248–251). Compare their relationships on the right and left sides. The sympathetic trunk (p. 252) lies posteriorly against the necks of the upper thoracic ribs and will be observed better in the posterior mediastinum. Find examples of **thoracic cardiac** nerves (p. 248), which derive from the vagus nerves and sympathetic trunks that course inferiorly on the superficial and deep surfaces of the aortic arch to form the cardiac plexus. From the left vagus nerve at the inferior surface of the aortic arch, identify the origin of the **left recurrent laryngeal nerve** (p. 248) as it passes deep and posterior to the ligamentum arteriosum. In the midline, note the deep positions of the **trachea** and **esophagus** and observe their relationships.

39. MIDDLE MEDIASTINUM AND PERICARDIUM

Review the boundaries the middle mediastinum and its contents. Complete the separation of the pericardium from the mediastinal pleura. Review the courses of the phrenic and vagus nerves and note their relationships to the roots of the lungs. Note the courses of the pericardiacophrenic vessels with the phrenic nerve between the peri-

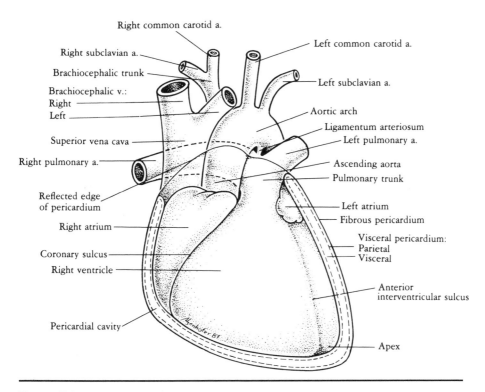

Right common carotid a.

Right subclavian a.

Brachiocephalic trunk

Brachiocephalic v.:
Right
Left

Superior vena cava

Right pulmonary a.

Reflected edge
of pericardium

Right atrium

Coronary sulcus

Right ventricle

Pericardial cavity

Left common carotid a.

Left subclavian a.

Aortic arch

Ligamentum arteriosum

Left pulmonary a.

Ascending aorta

Pulmonary trunk

Left atrium

Fibrous pericardium

Visceral pericardium:
Parietal
Visceral

Anterior
interventricular sulcus

Apex

Figure 5.6

cardium and the pleura (p. 243). Trace these vessels and nerve to the diaphragm.

The **pericardium** (Figure 5.6), a serous sac that surrounds the heart, extends from the sternal angle superiorly to the diaphragm inferiorly, where it fuses to the central tendon of the diaphragm (pp. 240–242; 249). Anteriorly, the pericardium is attached to the sternum by the superior and inferior sternopericardial ligaments. These ligaments were broken when the chest wall was removed. Review the layers of the pericardium, noting that there are two components: **fibrous** and **serous** (Figure 5.6). The fibrous coat, the most external layer of the pericardium, is in contact with the pleura and sternum and is fused to the diaphragm and continuous with the great vessels at the sternal angle. Open the pericardium by an incision that begins at the ascending aorta and continues inferiorly and to the left to the apex of the heart. Place your hand in the opened space and explore the **pericardial cavity.** This cavity is lined by two components of the serous pericardium: externally by the **parietal layer of the serous pericardium** (fused to the inner aspect of the fibrous pericardium) and internally by the **visceral layer of the serous pericardium** (fused to the epicardial surface of the heart). Within the pericardial cavity, explore the following two sinuses. The **transverse sinus** (pp. 243; 246) is located horizontally at the base of the heart and is found by inserting two fingers between the ascending aorta and pulmonary trunk anteriorly and the pulmonary veins and superior vena cava posteriorly. The **oblique sinus** (p. 246) is located posteriorly between the right and the left pulmonary veins on the diaphragmatic surface of the heart. To locate the oblique sinus, elevate the apex of the heart and insert your fingers posterior to the heart into the blind cul-de-sac between the right and the left pulmonary veins.

The great vessels (Figure 5.6) at the base of the heart (superior vena cava, ascending aorta and pulmonary trunk) pierce the pericardium to reach the heart. The reflection from visceral to parietal pericardium occurs across the ascending aorta, superior vena cava and pulmonary trunk at the sternal angle.

40. EXTERNAL FEATURES OF THE HEART

With the pericardium opened, study and define the surfaces and borders of the heart before continuing the dissection (pp. 228–230; 234). Note and outline their relationships to the anterior chest wall, diaphragm, lungs and structures within the posterior mediastinum. With the heart *in situ,* observe the **base,** which is located superiorly at the great vessels, the **apex,** which is inferiorly and to the left and formed by the left ventricle, and the **anterior,** or **sternocostal,** surface, which is facing you (pp. 228; 242–244). Identify the **right border,** formed by the right atrium, and the **left border,** formed mainly by the left ventricle. A sharp **inferior margin** separates the sternocostal surface from the **diaphragmatic surface,** which rests against the diaphragm. The **posterior surface,** composed of the left and right atria, can be observed best after the heart is removed.

Remove the heart to facilitate further dissection by cutting through the midpoint of the great vessels of the heart just inferior to the sternal angle, leaving segments of the vessels on the cadaver and on the heart. Next, cut the inferior vena cava at the point where it penetrates the diaphragm inferiorly. Lift the apex of the heart and cut the right and left pulmonary veins on each side at the points where they pierce the pericardium. The heart can now be lifted and removed from the body.

Review the three surfaces of the heart described previously while maintaining their anatomic relationships (p. 228). Identify and observe the courses of the **coronary** and **anterior** and **posterior interventricular sulci** (Figure 5.6), which are found on the surfaces of the heart (pp. 228–229; 234; 236; 242). Relate these sulci to the partitioning of the internal chambers of the heart. Remove the epicardial fat, which is lodged within the sulci, being careful to avoid damaging the coronary arteries within them.

41. CORONARY ARTERIES

Identify the **right coronary artery** in the right terminus of the coronary sulcus. The right coronary artery arises from the right aortic sinus of the ascending aorta and initially runs between the pulmonary trunk and the right auricle (pp. 231; 236; 244). Lift the right auricle to observe this part of the artery. It continues inferiorly in the coronary sulcus toward the inferior margin between the right atrium and the right ventricle on the sternocostal surface. From this anterior course of the right coronary artery, attempt to identify the small but important **sinoatrial nodal artery,** which is one of the first branches from the right side of the artery close to its origin. This nodal artery encircles the superior vena cava posteriorly to supply the sinoatrial node. A **conal** branch arises from the left side of the right coronary artery. Also, note the numerous branches that supply the sternocostal surfaces of the right atrium and right ventricle. Leaving the sternocostal surface, the right coronary artery continues its course posteriorly in the coronary sulcus between the right ventricle and the right atrium. A marginal branch to the anterior surface of the right ventricle is usually formed as the artery passes around the inferior margin. On reaching the superior end of the posterior interventricular sulcus, the right coronary artery turns inferiorly into the posterior interventricular sulcus and becomes the **posterior interventricular artery** (p. 236). This artery courses inferiorly in the posterior interventricular sulcus on the diaphragmatic surface of the heart between the two ventricles. The posterior interventricular artery supplies branches to both ventricles and numerous branches that penetrate the posterior third of the interventricular septum. It then anastomoses at the apex with the anterior interventricular branch of the left coronary artery. At the point of origin of the posterior interventricular artery, there are anastomotic branches that pass in the coronary sulcus across the midline to join the circumflex artery, a branch of the left coronary artery. Also at the same position, note the origin of the small but important **atrioventricular nodal artery,** which penetrates deeply through the interatrial septum to supply the atrioventricular node.

The **left coronary artery** arises from the left aortic sinus and is located in the left terminus of the coronary sulcus (pp. 228; 231; 236). Follow the left coronary artery in the coronary sulcus to the left of the pulmonary trunk and deep to the left auricle. It runs a short distance and then divides into the **anterior interventricular** and **circumflex arteries.** The left coronary artery may give rise to the sinoatrial nodal artery. The anterior interventricular artery enters the anterior interventricular sulcus on the sternocostal surface of the heart and runs inferiorly to the apex between the right and the left ventricles. During its course, it supplies branches to the sternocostal surfaces of both ventricles (diagonal branches) and branches that penetrate the anterior two-thirds of the interventricular septum. At the apex, the anterior interventricular artery anastomoses with terminal branches of the posterior interventricular artery. The circumflex artery continues a direct course in the coronary sulcus to the posterior surface of the heart between the left atrium and the left ventricle. Anteriorly, it supplies the left atrium and left ventricle and then gives rise to several obtuse and marginal branches to the left margin of the heart. In its posterior position in the coronary sulcus, the circumflex artery lies deep to the coronary sinus (see the next paragraph) and supplies the left atrium and left ventricle (**posterior artery of the left ventricle**).

Identify and follow the courses of the **great, middle** and **small cardiac veins** (pp. 236; 244). Note their relationships to the coronary arteries. Other cardiac veins include the posterior vein of the left ventricle on the posterior surface of the left ventricle and the anterior and smallest cardiac veins on the sternocostal surface (pp. 236; 244). Identify the **coronary sinus** in the posterior aspect of the coronary sulcus between the left atrium and the left ventricle. Determine which cardiac veins drain into the coronary sinus (p. 236). Note the relationships of these veins in the sulci to the courses of the coronary arteries.

42. INTERNAL FEATURES OF THE HEART

The remainder of the dissection involves studying the internal features of the heart and requires opening each of the four chambers.

A. Right Atrium

Open the right atrium anteriorly (Figure 5.7, incision 1) by an incision that begins superiorly at the base of the superior vena cava and continues inferiorly 1 cm to the right of the coronary sulcus to the inferior and right borders of the heart (pp. 232; 235; 245). Open the flap and note the smooth (**sinus venarum**) and muscular portions of the chamber, which are separated by the **terminal crest.** Identify the **pectinate muscles, opening of the superior vena cava, open-**

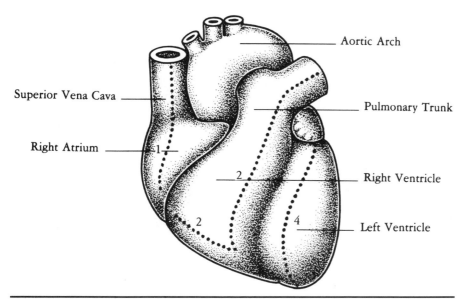

Superior Vena Cava

Right Atrium

Aortic Arch

Pulmonary Trunk

Right Ventricle

Left Ventricle

Figure 5.7

ing of the **inferior vena cava** and its **valve** and **opening of the coronary sinus** and its **valve,** which is located between the valve of the inferior vena cava and the right **atrioventricular aperture.** On the right side of the interatrial wall, identify the depressed **fossa ovalis** and surrounding elevated margin called the **limbus fossa ovalis.**

B. Right Ventricle

Open the right ventricle (Figure 5.7, incision 2) by an incision that begins at the cut end of the pulmonary trunk and continues inferiorly through its anterior wall and the anterior surface of the right ventricle to its inferior border. This cut should be made approximately 1 cm parallel to the right of the anterior interventricular sulcus. At the inferior end of this cut, continue the incision at a right angle toward the coronary sulcus and parallel to the inferior border. Stop this incision just short of the coronary sulcus. The anterior wall of the right ventricle can now be opened and the chamber can be cleaned of any blood (pp. 232; 235; 245).

Observe and identify within the right ventricle the **anterior, posterior** and **septal cusps of the right atrioventricular valve** (tricuspid), **anterior** and **posterior papillary muscles, trabeculae carneae, chordae tendineae, septomarginal trabecula** (moderator band) and **supraventricular crest.** Note that the upper portion

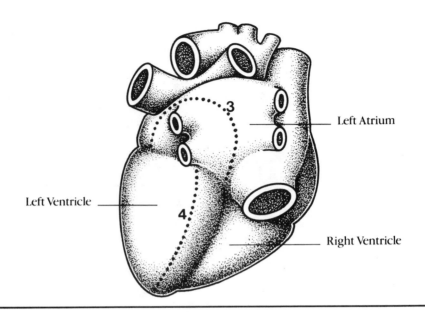

Left Atrium

Left Ventricle

Right Ventricle

Figure 5.8

of the right ventricular chamber tapers into a smooth-walled portion called the **conus arteriosus,** which leads to the **pulmonary trunk.** Identify the **right, left** and **anterior cusps of the pulmonary semi-lunar valves,** noting their structures and orientations (p. 233). The anterior wall and anterior papillary muscle of the right ventricle are related to the sternocostal surface of the heart and the inferior wall and posterior papillary muscle are related to the diaphragmatic surface. The medial wall (posteromedial in position) is formed by the interventricular septum, which bulges into the right ventricle.

C. Left Atrium

Open the left atrium (Figure 5.8, incision 3) by an arched incision superior to the coronary sinus between the right and the left pulmonary veins. Clear the cavity of blood and identify the **pectinate muscles, openings of the right** and **left pulmonary veins, left atrioventricular aperture, valvula foraminis ovalis** and **auricle.**

D. Left Ventricle

This last chamber can be opened (Figures 5.7 and 5.8, incision 4) by inserting a scalpel through the anterior left ventricular wall just inferior to the coronary sulcus and slightly to the left of the anterior interventricular sulcus. Continue the incision inferiorly and parallel to the anterior interventricular sulcus around the apex of the heart and then superiorly on the diaphragmatic surface to the left of the posterior interventricular sulcus as high as the coronary sulcus and sinus. Open this chamber and identify the **anterior** and **posterior papillary muscles, anterior** and **posterior cusps of the left atrioventricular valve** (mitral), **trabeculae carneae, chordae tendineae** and **interventricular septum** on the medial wall (pp. 230; 235). The anterior papillary muscle is attached to the sternocostal surface of the left ventricle and the posterior papillary muscle is attached to the diaphragmatic surface. Note that the size of the muscular structures and the thickness of the wall are greater in the left than in the right ventricle. Adjacent to the anterior cusp of the left atrioventricular valve is the entrance to the ascending aorta. This entrance is also related to the membranous part of the interventricular septum, which separates this part of the left ventricle from the lower part of the right atrium and upper part of the right ventricle.

Holding the heart in the anatomical position, observe the cut end of the ascending aorta and pulmonary trunk and identify the **right, left** and **posterior** (noncoronary) **cusps of the aortic semilunar valve** (p. 233) and the **right, left** and **anterior cusps of the pulmonary semilunar valve.** Note their structures and orientations and identify the ostia of the **right** and **left coronary arteries** on the superior (or aortic) side of the aortic valves.

43. CONDUCTING SYSTEM OF THE HEART

The conducting system consists of specialized cardiac muscle fibers that are responsible for initiating and controlling the normal cardiac cycle. This system is under the control of the sympathetic and parasympathetic divisions of the autonomic nervous system. Dissection of the conducting system is difficult in human cadavers, but students should be aware of its location and distribution within the heart. The sinoatrial node (pacemaker) initiates the contraction and is located in the anterior right atrial wall at the upper end of the terminal sulcus at the base of the superior vena cava (p. 235). From the sinoatrial node, the electrical impulse spreads through the muscle fibers of the left atrial wall to the atrioventricular node, which is located in the interatrial wall between the coronary sinus and the right atrioventricular opening. The atrioventricular node leads to the atrioventricular bundle, which passes to the left of the membranous part of the interventricular septum and divides into the right and left crura (bundle branches) (p. 235). Each crus descends beneath the endocardium on each side of the interventricular septum to reach the inferior end of the septum, where each crus turns superiorly to provide fibers (Purkinje fibers) to the remainder of the ventricles.

44. PULMONARY VESSELS AND MAIN BRONCHI

Remove the remaining posterior part of the pericardium to continue your dissection. Identify and follow the courses of the **pulmonary arteries** (Figure 5.9) and **veins** to the hila of the lungs (pp. 248–251). Also, identify the bifurcation of the **trachea** at the sternal angle into the **right** and **left main bronchi** (Figure 5.9) and follow them in the roots of the lungs (p. 249). Note the relationships of the pulmonary vessels and bronchi to each other, to the chambers of the heart and to the great vessels at the base of the heart. The right pulmonary artery and vein pass posterior to the right atrium to reach the lung. The right main bronchus is larger, shorter and more vertical than the left main bronchus, which passes deep to the aortic arch and anterior to the esophagus and descending thoracic aorta.

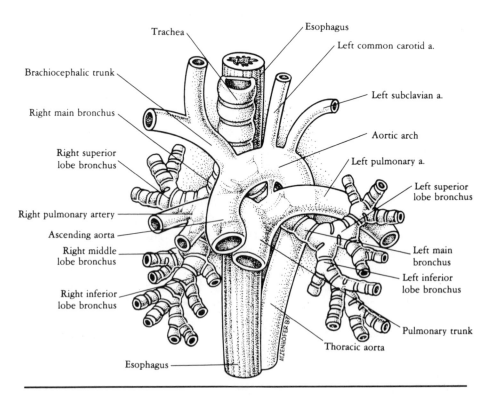

Figure 5.9

45. LUNGS

Review the relationships of the parietal and visceral pleurae to the lungs (p. 244). Identify the surface projections of the lungs to the chest wall and compare them to the surface projections of the parietal pleura (p. 224). Inferior to the roots of the lungs, the reflection of mediastinal pleura forms a double-layered pulmonary ligament that drops inferiorly along the mediastinal surfaces of the lungs. Open more completely the pleural sac and remove both lungs by cutting through all structures that enter and leave the lungs at the hila (pp. 248–250). Be careful to preserve the phrenic and vagus nerves. Identify the **bases** and **apices** of the lungs and their **costal, cervical, diaphragmatic** and **mediastinal surfaces** (Figure 5.10). Note that the margins of the lungs are sharp, especially at their inferior borders. Separate and identify the **lobes** of the lungs (superior, middle and inferior lobes of the right lung and superior and inferior lobes of the left lung) and identify the **oblique fissure** on both lungs and the **horizontal fissure** on the right lung (Figure 5.10). Observe the relationships of the cut structures at the roots of the lungs, especially the pulmonary arteries, veins and main bronchi (Figure 5.10) (pp. 225; 254–255). Dissect superficially into the hilum of each lung and identify the division of the main bronchi into secondary (lobe) bronchi,

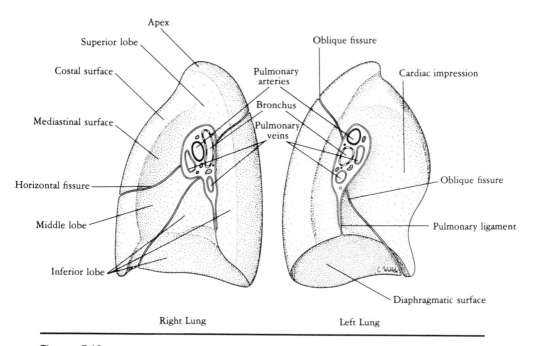

Right Lung   Left Lung

Figure 5.10

three in the right lung and two in the left lung. In the cadaver, the relationships of many structures within the mediastinum to the lungs can be determined by identifying the impressions made by these structures on the mediastinal surfaces of the lungs after death (p. 225). Identify as many of these structures as possible by use of an illustration from your textbook and atlas.

46. BRONCHOPULMONARY SEGMENTS

For a more detailed study of the lungs, you can now follow the distributions of the pulmonary vessels and lobe bronchi more deeply into the hilum of each lung to demonstrate their segmental supply to units of lung tissue called **bronchopulmonary segments** (pp. 226–227). Dissect into the hila and identify the division of the secondary (lobe) bronchi into tertiary bronchi by use of the terminology in your textbook. Also, note the branching of the pulmonary arteries and veins. The pulmonary veins are intersegmental in their drainage, whereas the bronchi and pulmonary arteries supply only one segment. Use your textbook to appreciate the bronchopulmonary segments projected on the surface of the lungs.

47. REVIEW OF MEDIASTINAL RELATIONSHIPS

Review the course of the **trachea** through the superior mediastinum to its bifurcation at the sternal angle into the **right** and **left main bronchi** and note the posterior position of the **esophagus** (p. 248). Review the relationships of the main bronchi to the **pulmonary vessels**. Note the **cardiac branches** of the sympathetic trunk and vagus nerves and their distributions through the superior mediastinum superficial and deep to the aortic arch to form the cardiac plexus (p. 248). Lateral extensions of these nerve fibers along the pulmonary arteries form the pulmonary plexus to the lungs. Review the courses of the **phrenic** and **vagus nerves** through the superior mediastinum and their entrances to the middle and posterior mediastina, respectively. Note their relationships to the trachea, esophagus and aortic arch (pp. 246; 248).

48. POSTERIOR MEDIASTINUM

Review the contents and identify the boundaries of the posterior mediastinum. Note the anterior relationship of the pericardium to the posterior mediastinum (p. 247) and remove any remaining parts of the pericardium to dissect the posterior mediastinum. In addition, it is *necessary* to elevate and strip away any remaining parietal

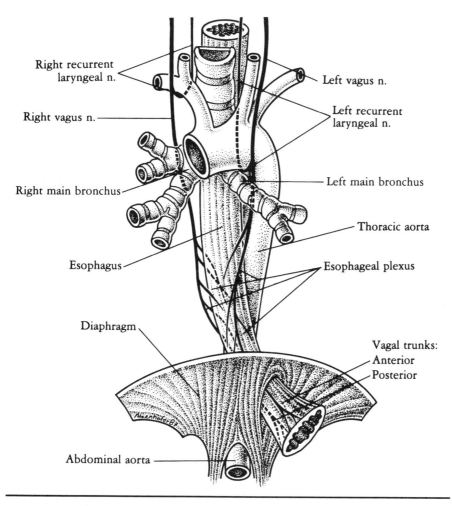

Figure 5.11

pleura that covers the posterior chest wall and vertebrae to proceed with the dissections that follow.

Clean and expose the **esophagus** (Figure 5.11) as it passes deep to the bifurcation of the trachea to enter the posterior mediastinum (pp. 248–253). In the posterior mediastinum, the esophagus is near the midline, in contact anteriorly with the posterior aspect of the pericardium and left atrium with the descending thoracic aorta to its left (pp. 246–246). As the esophagus descends anterior to the vertebral bodies, it inclines to the left and passes anterior to the thoracic aorta at about the level of the eighth thoracic vertebra. The esophagus leaves the thorax by passing through the **esophageal hiatus** of the diaphragm with the vagus nerve trunks at the level of the tenth thoracic vertebra.

The courses of the two **vagus nerves** (Figure 5.11) in the posterior mediastinum are closely related to the esophagus (pp. 248–251). After passing the level of the sternal angle, the vagus nerves enter the posterior mediastinum and pass posterior to the roots of the lungs to reach the surfaces of the esophagus. The left vagus nerve passes primarily to the anterior surface of the esophagus and the right vagus nerve passes mostly to the posterior esophageal surface. On the esophagus, the vagus nerves form the **esophageal plexus** (p. 249), which inferiorly consists of two large trunks, the **anterior** and **posterior vagal trunks.** These trunks pass through the diaphragm with the esophagus and distribute to the viscera in the abdomen.

Next, identify the **descending thoracic aorta** (pp. 250–253; 255) in the posterior mediastinum (Figure 5.11). It begins on the left side of the vertebral column at the level of the intervertebral disc between the fourth and fifth thoracic vertebrae, where it is continuous superiorly with the aortic arch. Identify as many of the following branches of this vessel as possible: **bronchial, esophageal, posterior intercostal, subcostal** and **superior phrenic arteries.** The latter two arteries are difficult to find. Review the relationship of the thoracic aorta to the esophagus and note the course of the artery through the aortic hiatus of the diaphragm (with the thoracic duct).

Venous drainage within the posterior mediastinum is accomplished primarily by the **azygos system** (pp. 251; 253–254). The **azygos vein** (Figure 5.12) is formed on the right posterior abdominal wall just inferior to the level of the diaphragm by the junction of the ascending lumbar and right subcostal veins. The azygos vein enters the thorax and can now be traced superiorly on the right side of the aorta to its termination as the **azygos arch** into the **superior vena cava.** Identify examples of the **right posterior intercostal veins** and the **right superior intercostal vein.** Elevate the aorta and note posteriorly the courses of the **hemiazygos** and **accessory hemiazygos veins** on the left receiving blood from the **left posterior intercostal veins.** Various parts of this venous system are often absent and numerous variations can occur. It is important to note that the azygos system provides collateral flow between the superior and the inferior venae cavae. Identify the **left superior intercostal vein** at the superior end of the accessory hemiazygos vein. This vein drains into the left brachiocephalic vein. Note that it separates the left vagus and left phrenic nerves at the left surface of the aortic arch.

The **thoracic duct** (pp. 22; 251) is the large, terminal lymphatic vessel that drains all the lymph of the body inferior to the diaphragm and the left side of the body superior to the diaphragm (Figure 5.12). The duct begins in the abdomen and enters the thorax by passing through the aortic hiatus of the diaphragm. The duct ascends poste-

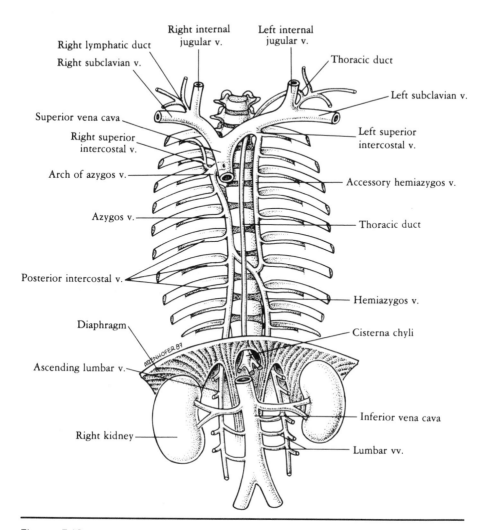

Figure 5.12

rior and to the right of the aorta between the aorta and the azygos vein. Follow this relationship superiorly to the level of the fifth thoracic vertebra, where the duct passes to the left posterior to the aorta and esophagus to enter the superior mediastinum. The duct then ascends into the root of the neck, where it terminates at the junction of the left subclavian and internal jugular veins. This termination of the duct can be better dissected in the root of the neck.

Review the courses of the **posterior intercostal vessels** and **intercostal nerves** in the posterior aspect of the intercostal spaces (pp. 254–255). Follow them proximally to the origins of the nerves and arteries and the terminations of the veins. The posterior intercostal arteries usually give rise to collateral intercostal arteries at the midaxillary line and then connect to the anterior intercostal arteries within the intercostal spaces, thus establishing collateral circulation between the aorta and the subclavian arteries via the internal thoracic artery. You will have to strip away remaining layers of the parietal pleura from the posterior and lateral chest walls to view the intercostal spaces.

Finally, identify the thoracic course of the **sympathetic trunk** (pp. 253–255), which can be observed after the pleura is removed. Review its course through the superior mediastinum. Initially in the posterior mediastinum, it passes anterior to the neck of the ribs at the lateral sides of the vertebrae. In the inferior part of the posterior mediastinum, the trunk courses onto the ventral surfaces of the vertebrae, where it passes posterior to the diaphragm and enters the abdomen. Note the positions of the thoracic **sympathetic chain ganglia,** which are located segmentally along the sympathetic trunk. From each ganglion, identify the **gray** and **white rami communicantes,** which connect to each intercostal nerve, the white rami being more lateral. Three **splanchnic nerves** arise from the thoracic chain ganglia and pass ventral to the vertebrae to enter the thorax through the crura of the diaphragm: the greater splanchnic nerve arises from the fifth to ninth ganglia, the lesser splanchnic nerve arises from the tenth and eleventh ganglia, and the least splanchnic nerve arises from the twelfth thoracic ganglion. The lesser and least splanchnic nerves are deep and posterior to the diaphragm and are difficult to expose at this time. The **greater splanchnic nerve** should be identified. Review the preganglionic and postganglionic distributions of the fibers in the thoracic part of the autonomic nervous system and understand their innervation of various visceral structures.

CHAPTER SIX

ABDOMEN

49. SUPERFICIAL STRUCTURES AND LANDMARKS

The abdomen is the part of the trunk that extends inferiorly from the diaphragm to the pelvis, where it becomes continuous with the pelvic cavity. The upper part of the abdominal cavity is protected by the lower thoracic wall and the lower part by the bones of the greater pelvis. However, most of the anterior and lateral aspects of the abdominal wall are formed by muscle layers. On the abdominal wall and skeleton, locate the following landmarks: the **infrasternal angle, costal margin** (formed by fusion of the seventh to tenth ribs), **iliac crest, anterior superior iliac spine, symphysis pubis, inguinal ligament** and **pubic crest** and **tubercle** (p. 175). Locate the two vertical bulges of the **rectus abdominis muscle** flanking each side of the midline. Also, identify the single, midline **linea alba,** which separates the two rectus muscles and two **lineae semilunares,** which border each muscle laterally. From your textbook and atlas, understand the separation of the abdominal wall into quadrants by superimposed horizontal and transverse lines and planes. One of the most important planes is the **transpyloric plane** (p. 187), a key anatomic landmark. Because many surgical approaches are made through the muscular walls, it is important to appreciate the surface projections of the viscera on the surface of the body while they are being studied in the dissections that follow.

Use Figure 6.1 to make the following skin incisions. First, make a vertical, midline incision (A-B) from the infrasternal angle to the symphysis pubis. In the male, continue this line around the base of the penis and through the midline of the scrotum. In the female (Figure 6.2), make this cut along the lateral margins of each labium majus to the posterior commissure.

In both sexes, next make two lateral, transverse cuts from the infrasternal angle (A-C) and umbilicus (D-E). The two transverse cuts should be carried laterally to the midaxillary line on each side of the body. Next, if the lower limb has not been dissected, make a cut from the symphysis pubis parallel and inferior to the inguinal ligament superiorly as far as the anterior superior iliac spines (B-F). Reflect the abdominal skin flaps laterally.

The superficial fascia of the abdominal wall contains large amounts of fat deposits. Remove the superficial fascia inferiorly to the scrotum or labia and identify examples of the cutaneous nerves and vessels discussed in the text that follows. As the fascia is removed, note the specialization of the superficial fascia in the lower third of the anterior wall. In this area, there is a **superficial fatty layer** and a **deep membranous layer.** Observe the membranous layer on the undersurface of the fatty layer and understand its in-

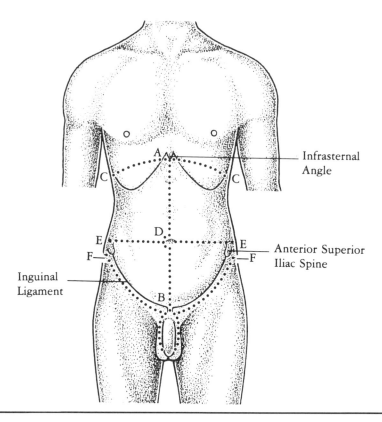

Infrasternal Angle

Anterior Superior Iliac Spine

Inguinal Ligament

Figure 6.1

ferior attachments to the iliac crest, fascia lata, inguinal ligament and ischiopubic rami and urogenital diaphragm of the perineum (see Chapter 7 for discussion). Deep to the membranous fascia, identify the communication from the abdominal wall into the scrotum or labia majora (abdominoscrotal or abdominolabial opening).

50. CUTANEOUS NERVES AND BLOOD VESSELS

The abdominal **cutaneous nerves** pass segmentally through the superficial fascia (pp. 188–189; 196). They are formed by the seventh to twelfth intercostal nerves with a contribution from the first lumbar nerve, which forms the **iliohypogastric nerves.** These cutaneous nerves supply the thoracic and abdominal walls. The seventh nerve is at the level of the xiphoid process, the tenth nerve distributes around the umbilicus, and the iliohypogastric nerve (pp. 194–196) is suprapubic. The **ilioinguinal nerve** (pp. 194–196) exits the superficial inguinal ring to enter the scrotum or labia. Identify examples of many of these nerves during removal of the superficial fascia.

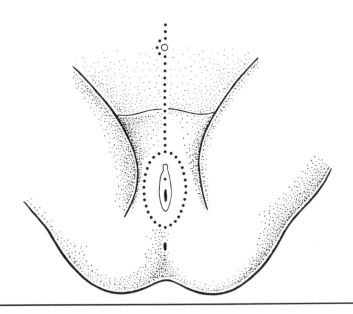

Figure 6.2

Within the superficial fascia inferior to the umbilicus, three sets of **cutaneous vessels** supply the abdominal wall (superficial epigastric, superficial circumflex iliac and superficial external pudendal) (pp. 187–188). These arteries are branches of the femoral artery in the thigh and the veins are tributaries of the greater saphenous vein of the thigh. Other aspects of the abdominal wall receive blood supply from branches of the intercostal, lumbar, musculophrenic arteries and the superior and inferior epigastric arteries. Many of these vessels may be difficult to dissect and you need not identify each of them in the cadaver. The veins of the adbominal wall form a collateral connection between the two venae cavae.

51. ANTEROLATERAL MUSCLES AND FASCIA

On each side, the anterior and lateral aspects of the abdominal wall are formed primarily by three layers of flat muscles and by the vertical rectus abdominis muscle in the midline. The three flat muscles originate laterally and insert medially in the rectus sheath and inferiorly in the inguinal region. These three flat muscles have alternating fiber directions, similar to those of the thoracic intercostal muscles, and function in support, compression, movement of the trunk, respiration, micturition, defecation and parturition. Use your textbook to review the origins, insertions, innervations and actions of each muscle described in the text that follows during its dissection.

Note that the lowest fibers of all three flat muscles insert in the inguinal region and will be left *intact* until the superior fibers of each layer have been identified. The inferior fibers of each muscle are related primarily to the inguinal region, which will be dissected later.

A. External Abdominal Oblique Muscle

Clean the surface of the **external abdominal oblique** muscle (pp. 188; 192), noting the directions and lateral origins of the fibers from the inferior eight ribs. The lowest fibers that attach to the iliac crest become almost vertical. As the upper fibers pass medially, they become aponeurotic and contribute to the anterior layer of the **rectus sheath,** with final insertion at the **linea alba** at the midline.

In the midaxillary line, carefully separate a window in the fibers of the external abdominal oblique muscle with scissors and identify the fascial plane between this muscle and the deeper **internal abdominal oblique** muscle (pp. 192; 194–195). The latter muscle can be identified by its different fiber directions. After this plane has been identified, enlarge the opening and complete the separation of the two muscles by making a vertical incision in the external oblique muscle between the iliac crest and the costal margin at the midaxillary line. Note that you cannot explore any further medially than the semilunar line, where the aponeurotic fibers of the external oblique muscle contribute to the rectus sheath. Next, make a horizontal incision in the external oblique muscle between the anterior superior iliac spine and the lateral border of the rectus adominis (Figure 6.3, A-B). At this time, do not dissect inferior to this line in the inguinal region. Superiorly at the ribs, separate and make an incision in the external oblique muscle between its attachment to the costal margin and the semilunar line of the rectus sheath (Figure 6.3, C-D). Pull the external oblique muscle medially to expose the **internal abdominal oblique** muscle (Figure 6.4).

B. Internal Abdominal Oblique Muscle

The fibers of the **internal abdominal oblique** muscle (pp. 192; 194–195) originate laterally from the thoracolumbar fascia, anterior two-thirds of the iliac crest and lateral two-thirds of the inguinal ligament. Note that the internal fibers pass superiorly and at a 90-degree angle to the external oblique muscle. Most of the fleshy muscle fibers become aponeurotic and contribute to the rectus sheath at the semilunar line to insert at the linea alba. The more-inferior fibers that arise from the inguinal ligament fail to reach the rectus sheath,

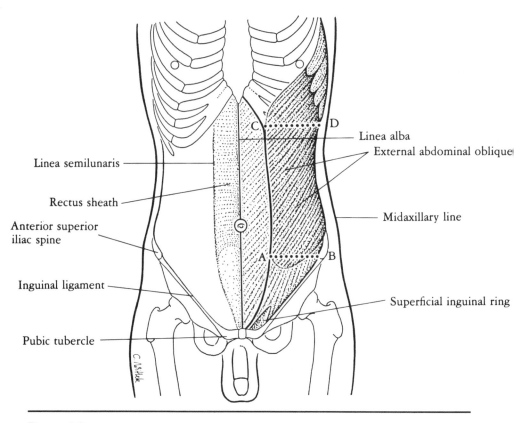

Figure 6.3

instead arching and turning downward to attach to the superior border of the symphysis pubis in the inguinal region (with the fibers of the transversus abdominis muscle). These inferior fibers are observed better during the dissection of the inguinal region.

The internal abdominal oblique muscle will now be reflected to identify the transversus abdominis muscle (p. 196). Clean the surface of the internal oblique muscle. Again with scissors, make a window in the fibers of the internal oblique muscle at the midaxillary line and with your fingers blunt dissect the fascial plane between the internal and the deeper transversus abdominis muscles. The internal oblique muscle is often fused to the transversus muscle. Note the difference in the directions of the fibers of the internal and transversus muscles. Make a vertical cut in the internal fibers at the midaxillary line between the iliac crest and the costal margin. Next, make similar horizontal cuts through the fibers of the internal oblique muscle, first from the anterior superior iliac spine to the rectus abdominis (Figure 6.4, A-B) and then at the infrasternal angle (Figure 6.4, C-D). Separate the internal fibers and pull the muscle medially.

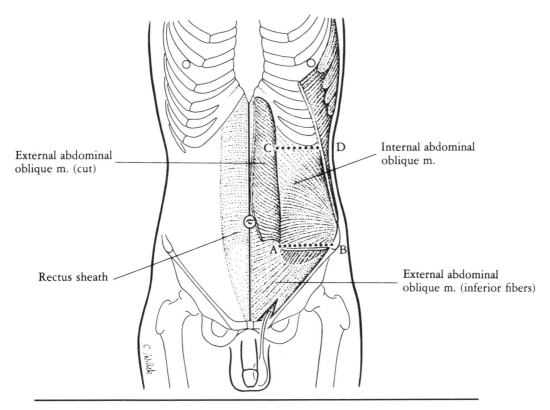

External abdominal oblique m. (cut)

Internal abdominal oblique m.

Rectus sheath

External abdominal oblique m. (inferior fibers)

Figure 6.4

Again, do not dissect inferior to the level of the anterior superior iliac spine into the inguinal region at this time.

C. Transversus Abdominis Muscle

The deepest layer of the flat muscles is the **transversus abdominis muscle** (p. 196), which is oriented transversely to the direction of the other two muscles (Figure 6.5). Note this muscle deep to the reflected internal oblique muscle. Review its extensive origin, noting the inferior fibers that arise from the lateral third of the inguinal ligament. Most of its superior aponeurotic fibers contribute to the rectus sheath, as do those of the other flat muscles. The fibers that arise from the inguinal ligament arch medially and inferiorly to the inguinal region and do not contribute to the rectus sheath. This insertion is in common with the inferior fibers of the internal oblique muscle and will be observed in the inguinal region. Note that most of the nerves and vessels that supply the abdominal wall and muscles pass forward in the fascial plane between the internal oblique and

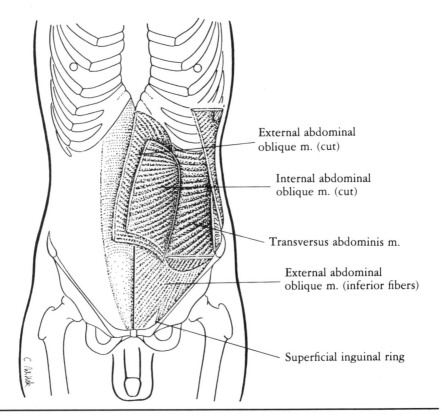

External abdominal
oblique m. (cut)

Internal abdominal
oblique m. (cut)

Transversus abdominis m.

External abdominal
oblique m. (inferior fibers)

Superficial inguinal ring

Figure 6.5

the transversus abdominis muscles (p. 196). Do not reflect the transversus abdominis muscle.

D. Rectus Abdominis Muscle and Sheath

Open the **rectus sheath** on each side of the linea alba with a long vertical incision at the middle of each sheath (pp. 192–196). Identify the fibers of the **rectus abdominis** on either side of the linea alba and review its superior and inferior attachments. **Transverse tendinous intersections** divide the muscle horizontally into segments and are fused to the rectus sheath. Cut the muscle transversely at its middle and identify the **superior** and **inferior epigastric vessels** deeply and follow their superior and inferior courses and note their relationships (pp. 194–195). Review and identify the layers of the **rectus sheath** and describe the pattern of contribution to it by the aponeuroses of the three flat muscles. Identify the **arcuate line** and describe the change in the structures that form the posterior layer of the rectus sheath at the arcuate line.

Deep to the abdominal wall muscles is a thin layer of transversalis fascia that lines the abdominal wall and cavity. This fascia separates the flat muscles from the deeper extraperitoneal connective tissue and parietal peritoneum. Understand the extent of the transversalis fascia from your textbook and its continuation as a common fascial layer that lines the thorax and abdominopelvic cavity.

52. INGUINAL REGION AND CANAL

The inguinal region is the inferior aspect of the anterior abdominal wall and contains the **inguinal canal.** The inguinal region is inferior to the horizontal cuts made earlier in the flat muscles between the anterior superior iliac spine and the linea semilunaris. The testes and spermatic cord pass through the canal during their descent into the scrotum. The round ligament occupies the same definitive position in the female. The inguinal canal is a weak area, vulnerable to herniation of the abdominal viscera through the anterior wall. The aponeurotic fibers of the inferior parts of the three flat muscles form the walls of the inguinal canal.

The superficial structures of the inguinal region are specializations of the **external abdominal oblique aponeurosis.** Identify the **superficial inguinal ring** (Figure 6.6, A), which appears as a triangular gap in the lowest aspect of the external abdominal oblique aponeurosis superior to the pubic tubercle (pp. 193; 197). The ring is bordered by **medial** and **lateral crura** that attach to the pubic crest and tubercle, respectively. In the superficial ring (p. 199), identify the **spermatic cord** in the male and **round ligament** in the female. Observe the ilioinguinal nerve (p. 197) on the surface of the spermatic cord (or round ligament) at the superficial ring and follow it into the scrotum or labia majora. The superficial ring is the external end of the inguinal canal and is an opening in the anterior wall for passage of potential hernias. As the spermatic cord exits the superficial ring, it carries with it an **external spermatic fascia** derived from the external oblique aponeurosis.

Identify the **inguinal ligament** (Figure 6.6, A), which extends between the **pubic tubercle** and the **anterior superior iliac spine** (pp. 197–198). This ligament is the inferior, recurred margin of the external oblique aponeurosis. Medially at the pubic tubercle, elevate the spermatic cord (round ligament) and note that the fibers of the inguinal ligament flatten and extend deeply and horizontally to form the **lacunar ligament.** The lacunar ligament then attaches to the pectineal line of the pubis and forms the **pectineal ligament.** The floor of the inguinal canal is formed by the lacunar ligament medially and the inguinal ligament laterally.

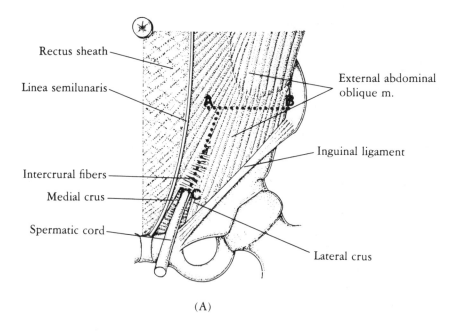

Rectus sheath

Linea semilunaris

External abdominal
oblique m.

Inguinal ligament

Intercrural fibers

Medial crus

Spermatic cord

Lateral crus

(A)

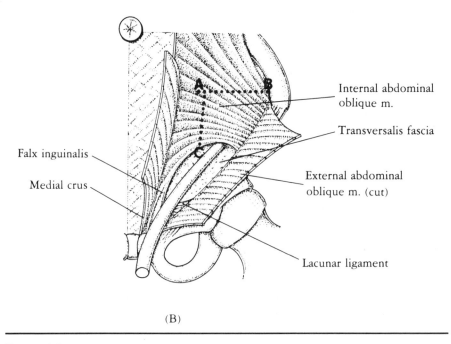

Internal abdominal
oblique m.

Transversalis fascia

Falx inguinalis

Medial crus

External abdominal
oblique m. (cut)

Lacunar ligament

(B)

Figure 6.6

To dissect the deeper relationships of the inguinal canal, make a vertical cut from the apex of the superficial inguinal ring superiorly through the external abdominal oblique muscle (Figure 6.6, A, A-C). Continue this cut superiorly to meet the horizontal cut already made through the external oblique muscle from the anterior superior iliac spine (Figure 6.4).

Reflect this triangular segment of the external oblique muscle inferiorly and laterally to open the anterior wall of the inguinal canal (Figure 6.6, B). The canal is about 4 cm in length and runs parallel to the inguinal ligament between the superficial ring medially and the deep inguinal ring laterally (p. 199). By following the spermatic cord (or round ligament) laterally, the cord can be seen passing through the **deep inguinal ring** (Figure 6.7). This deep ring is located superior to the inguinal ligament about half the distance between the symphysis pubis and the anterior superior iliac spine. To better display the deep ring on one side of the body, make a vertical cut through the arching fibers of the internal oblique muscle (Figure 6.6, B, A–C), creating a triangular flap that can be reflected inferiorly similar to that of the external oblique muscle. This vertical cut will meet the horizontal cut made previously in the internal oblique muscle. Separate this flap from the deeper fibers of the transversus abdominis (Figure 6.7, A). The common insertion of the internal oblique and transversus abdominis muscles to the symphysis pubis forms the **falx inguinalis** (conjoint tendon), which is deep to the medial crus. The inferior fibers of the internal oblique muscle that originate from the inguinal ligament (p. 197) make an additional contribution to the anterior wall of the canal, which is formed primarily by the external oblique aponeurosis. The **floor,** formed by the inguinal and lacunar ligaments, has also been described. The **roof** is formed by the fibers of the internal oblique and transversus muscles, which arise from the inguinal ligament and arch over the inguinal canal to insert posteriorly at the pubic crest. As was noted previously, the parallel courses of these two muscles to their common insertion posterior to the medial crus of the superficial ring form the falx inguinalis (or conjoint tendon). The **cremaster muscle** and **fascia** of the spermatic cord are derived from the arching fibers of the internal oblique muscle (p. 197). The **posterior wall** of the canal is formed medially by the falx inguinalis, which adds considerable strength to this part of the posterior wall. Laterally, the posterior wall is strengthened by the positions and courses of the **inferior epigastric artery** and **vein** (p. 198). Expose the origin of this artery from the external iliac artery as it passes deep to the inguinal ligament by carefully removing the fascia at the deep ring. Then, remove the investing fascia of the artery and follow the artery superiorly

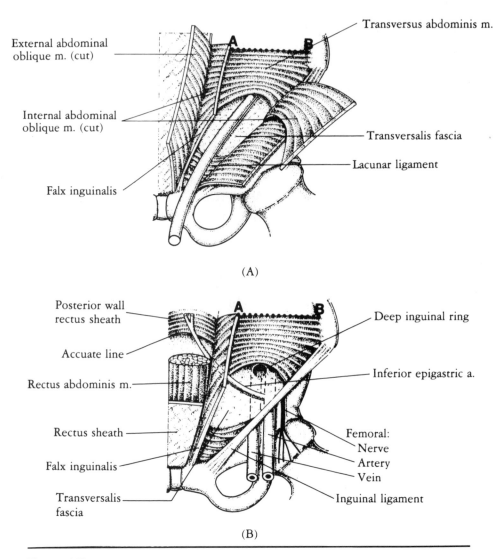

Figure 6.7

through the transversalis fascia into the rectus sheath (Figure 6.7, B). Note that the deep inguinal ring is immediately lateral to the artery. The posterior wall between the inferior epigastric artery and the falx inguinalis is called the weak area and is formed primarily by the transversalis fascia and peritoneum. This weak area is the region of the posterior wall through which direct hernias occur; indirect hernias follow the course of the spermatic cord (round ligament) through the deep inguinal ring lateral to the inferior epigastric artery. The deep ring is a dimple in the transversalis fascia. The inter-

nal spermatic fascia that covers the spermatic cord is a continuation
of the transversalis fascia.

53. SPERMATIC CORD

Review the origin and arrangement of the layers of the spermatic
fascia and cremaster muscle that cover the spermatic cord. Identify
the **ilioinguinal** and **iliohypogastric nerves,** noting their courses
and relationships in the canal (pp. 197–199). Remove these layers of
the cord, noting the **cremaster,** and identify the **ductus deferens,
deferential artery, testicular artery** and **pampiniform plexus of
veins** (pp. 198–199). Follow these structures to the testis in the
scrotum by separating the scrotal walls and septum. Identify the
epididymis.

54. PERITONEUM

To study the peritoneum and visceral contents of the abdominal cavity, open the anterior abdominal wall, as shown in Figure 6.8. Return
the dissected layers of the wall to their positions and begin the incision at the xiphoid process. Continue the incision inferiorly along the
left aspect of the linea alba and around the left side of the umbilicus
and then inferiorly to the symphysis pubis. While making this incision, elevate the anterior wall and use your hand to separate the
abdominal wall from the peritoneum (often fused) and abdominal
contents deep to the cut to avoid damaging them. Superiorly, cut the
fibers of the rectus abdominis from their attachments to the costal
margin several centimeters on each side of the xiphoid process. Next,
immediately inferior to the umbilicus, make two horizontal cuts
through the abdominal wall toward the anterior superior iliac spine.
These horizontal cuts were partially made during prior dissection of
the inguinal region. Preserve as many as possible of the relationships
of the muscle layers of the abdominal wall that were previously dissected. The four triangular flaps produced by these new cuts can now
be opened to view the abdominal contents.

The peritoneum is a complex invaginated, serous membrane that
invests many of the abdominal viscera, blood vessels and nerves that
will be described. Before removing or disturbing any of the normal
relationships, study the general arrangement of the peritoneum
along with the *in situ* positions of the viscera (pp. 265; 280–281).
Deep to the reflected muscular wall, make a vertical incision in the
peritoneum to expose the viscera. To examine the viscera, palpate
and blunt dissect only with the hands and fingers. Adhesions often

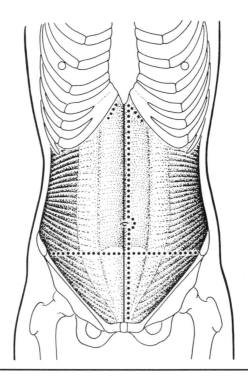

Figure 6.8

occur between viscera or between viscera and the body wall and any adhesion will have to be broken as you proceed.

Understand that the peritoneum is divided into parietal and visceral layers (pp. 265–266). The **parietal layer** lines the abdominal wall, pelvic wall and inferior surface of the diaphragm. The parietal peritoneum forms a closed sac by reflecting from the body wall over the surfaces of many viscera as a **visceral layer** that is fused to the surface of the organ. Viscera with such peritoneal coverings are referred to as intraperitoneal organs. Many of the visceral reflections form named mesenteries and ligaments that provide considerable mobility for the viscera. In fact, most of the vascular and nerve supply to the intraperitoneal viscera is distributed through these visceral reflections. This relationship between the vascular supply and the peritoneum is of primary importance. The two layers of the peritoneum are separated by a potential space, the peritoneal cavity, which contains a small layer of fluid that allows the viscera to move over each other without friction.

The retroperitoneal viscera are the organs posterior to the parietal peritoneum and thus are covered on only one surface by the peritoneum. The aorta, pancreas, ascending and descending colon, kidneys and inferior vena cava are examples of retroperitoneal structures on the posterior abdominal wall. A great deal of the dissection of the viscera, vessels and nerves will be accomplished simply by stripping away their peritoneal investment.

Initially, identify the **greater omentum** (p. 279), which attaches to the inferior, left border (greater curvature) of the stomach. This structure is a peritoneal apron formed by several layers of peritoneal reflections that contain considerable amounts of fat. It is suspended from the greater curvature of the stomach and is divided into three parts: the **gastrolienal ligament,** which extends between the stomach and the spleen; the **gastrophrenic ligament,** which attaches to the diaphragm; and the **gastrocolic ligament,** which extends to the transverse colon, where it is fused deeply to the **transverse mesocolon** (p. 279). The part of the greater omentum that extends inferior to the transverse colon varies in size considerably and may spread out over most of the viscera. Free the omentum from its adhesions to the viscera and body wall and observe its extent.

From the superior, right border (lesser curvature) of the stomach, identify the thin, transparent **lesser omentum** (p. 285), which extends from the stomach and first part of the duodenum to the visceral surface of the liver. Observe that the lesser omentum is formed by two ligaments, the **hepatogastric** (to the stomach) and **hepatoduodenal** (to the first part of the duodenum) ligaments that are continuous with each other.

Examine and explore the relationships of the following viscera (pp. 280–281) to the peritoneum before proceeding with any detailed dissection:

1. **Stomach.** Note its attachment to the greater and lesser omenta, as described previously.

2. **Liver.** In addition to the attachment of the lesser omentum to the stomach, note its attachment to the anterior body wall by the **falciform ligament.** The continuation of the falciform ligament on the liver forms the coronary ligaments, which will be examined later.

3. **Diaphragm.** Peritoneal reflections from the liver and stomach connect to the diaphragm.

4. **Gallbladder.** This organ projects from the inferior border of the liver and is attached to the visceral surface of the liver, which is covered by the peritoneum.

5. **Spleen (lien).** This organ lies against the diaphragm posterior to the left aspect of the stomach and is well protected by the inferior left costal margin. The spleen is connected to the stomach by the **gastrolienal ligament** and rests on a ligament that extends

from the splenic flexure to the diaphragm, the **phrenicocolic ligament** (sustentaculum lienis).

6. **Small intestines.** Only the coils of the **jejunum** and **ileum** can be identified at this time. Carefully pull these coils from their positions and identify the extensive **mesentery** that attaches them to the posterior abdominal wall.

7. **Large intestine.** This organ begins at the **cecum,** which is continuous with the ileum at the ileocecal valve. If present, identify the **appendix** attached to the cecum. Identify the **ascending, transverse, descending** and **sigmoid colon.** Noth that the transverse and sigmoid parts have peritoneal reflections, the **transverse mesocolon** and **sigmoid mesocolon,** respectively. The transverse and sigmoid colon can be freely elevated from the cavity. By contrast, the ascending and descending colon are retroperitoneal and should be left undisturbed at this time. They are completely covered anteriorly by the parietal peritoneum and are embedded in fusion fascia. Other viscera, such as the duodenum and pancreas, are also retroperitoneal and will be exposed later.

The peritoneal cavity is divided into two compartments, the **greater** and **lesser peritoneal sacs** (pp. 286–287). The lesser peritoneal sac (also called the **omental bursa**) is found posterior to the stomach and the greater and lesser omenta. The omental bursa communicates with the greater sac via the **epiploic foramen** (of Winslow) (pp. 286–287). With one or two fingers, identify this foramen posterior to the right free margin of the lesser omentum. The boundaries of the foramen are as follows: inferiorly, the first part of the duodenum; superiorly, the liver (caudate lobe); posteriorly, the retroperitoneal inferior vena cava; and anteriorly, the free margin of the lesser omentum. The omental bursa can be explored by making a transverse cut through the lesser omentum. In the posterior wall of the bursa, the pancreas, left kidney and left suprarenal gland can be palpated deep to the peritoneum (pp. 289–290).

55. SMALL AND LARGE INTESTINES

From their *in situ* positions in the abdominal cavity, pull the parts of the small intestine (the jejunum and ileum) from the cavity and identify and review their positions and extents (Figure 6.9). Elevate the **cecum** and **transverse** and **sigmoid colon** (Figure 6.9) (pp. 279–281). All these structures are intraperitoneal. The ascending and descending colon are retroperitoneal. Note the following peritoneal relationships: the **mesentery,** which invests the jejunum and ileum; the partial peritoneal reflection at the cecum; the **transverse**

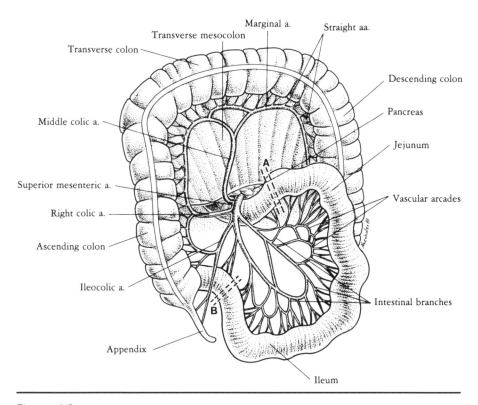

Figure 6.9

mesocolon, which invests the transverse colon; and the **sigmoid mesocolon,** which invests the sigmoid colon.

Locate the **duodenojejunal flexure** (Figure 6.9) at the proximal, fixed end of the jejunum at the left aspect of the second lumbar vertebra (p. 281). At this point, the duodenum turns anteriorly to become the jejunum, which enters the mesentery. Identify the ileocecal junction distally at the cecum. The root of the mesentery begins superiorly at the duodenojejunal junction and extends obliquely and inferiorly to the right for about 15 cm to reach the right iliac fossa at the cecum. This line represents the attachment of the mesentery to the posterior abdominal wall.

At the superior end of the root of the mesentery, palpate the bulge of the **superior mesenteric artery** and **vein** (pp. 275–277; 282–283), which enter the mesentery superior to the third part of the duodenum. Carefully tease away the peritoneum and identify and follow these vessels distally, noting their branches, which will be described (Figures 6.9 and 6.10). Observe that many autonomic nerves run with the vessels within the mesentery. Begin exposing the vessels proximally where they cross the third part of the duode-

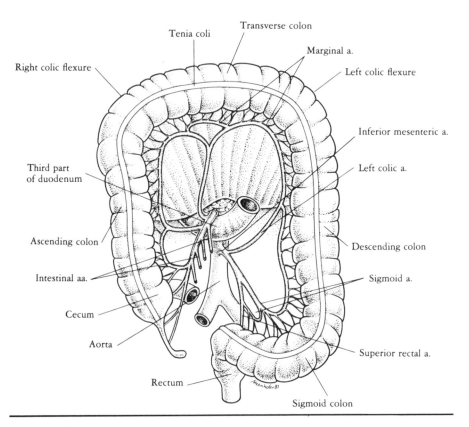

Figure 6.10

num to enter the mesentery. The superior mesenteric artery is the second of the three unpaired visceral branches of the abdominal aorta. It arises at the inferior aspect of the first lumbar vertebra immediately inferior to the celiac trunk. The initial course of this artery is deep to the neck of the pancreas, where arteries that supply the pancreas and duodenum are formed. The artery then crosses the third part of the duodenum anteriorly and enters the mesentery. Do not dissect the initial course deep to the pancreas at this time. The **superior mesenteric vein** can be seen anterior and to the right of the artery. This vein also ascends deep to the neck of the pancreas, where its termination joins the splenic vein to form the hepatic portal vein, which will be dissected later. To facilitate identification of the branches of the superior mesenteric vessels, carefully remove the peritoneum from the right posterior abdominal wall on the right aspect of the root of the mesentery between the ascending colon and the superior mesenteric vessels. Free the ascending colon from its retroperitoneal fascial bed. Be careful not to damage the blood vessels to the colon (described later), which are deep to the peritoneum. Identify and clean the following branches (Figure 6.9) of the superior mesenteric artery from the mesentery (pp. 275–277; 282–283):

1. **Intestinal branches to the jejunum and ileum.** These branches are represented by a number (12 to 20) of arteries that branch within the mesentery from the left aspect of the superior mesenteric artery. Follow several of these arteries distally to the surfaces of the jejunum and ileum. As these vessels reach the intestines, they form arches and loops with the adjacent artery to form rows of arterial arcades and straight arteries. Identify these vessels and note that they are more numerous at the ileum.

2. **Ileocolic artery.** This artery branches from the right side of the superior mesenteric artery to reach the ileocecal junction in the mesentery. The ileocolic artery provides branches to the distal ileum, cecum and appendix and forms an ascending branch that anastomoses with the right colic artery along the ascending colon.

3. **Right colic artery.** This artery passes posterior to the peritoneum in the fascia on the posterior body wall to reach the midpoint of the ascending colon, where it forms ascending and descending branches. This vessel can be highly variable and may be absent. It often branches from the ileocolic artery.

4. **Middle colic artery.** This artery branches from the superior mesenteric artery at the inferior border of the neck of the pancreas. It enters the transverse mesocolon, forming right and left branches on the transverse colon. The channel formed by the right, middle and left colic arteries follows the borders of the ascending, transverse and descending large colon and is called the **marginal artery.**

Next, remove the jejunum and ileum (Figure 6.9). At the duodeno-jejunal and ileocecal junctions, place two strings at each location around the intestine. At each location, cut between these two sets of strings and remove the small intestine by cutting its peritoneal attachments along the *surfaces* of the jejunum and ileum. The mesentery and its vessels are left intact and the only vessels severed are those that attach directly to the intestine (straight arteries).

The distributions of the inferior mesenteric vessels (Figure 6.10) can now be dissected (pp. 276–277; 284) on the left posterior abdominal wall. Examine the posterior abdominal wall on the left side, noting the splenic flexure, descending colon and sigmoid colon. Carefully remove the peritoneum that covers this area to free the descending colon from the retroperitoneal fascia. Do not destroy the vessels posterior to the peritoneum. The **inferior mesenteric artery** is the most inferior of the three unpaired visceral branches of the abdominal aorta. It arises from the aorta at the level of the third lumbar vertebra, 3 or 4 cm superior to the aortic bifurcation. As the peritoneum is being removed, identify the inferior mesenteric artery, which branches from the aorta, and follow its branches distally. The following branches should be identified:

1. **Left colic artery.** This artery is the most superior branch and passes posterior to the peritoneum to reach the descending colon. It forms ascending branches that contribute to the marginal artery and descending branches that meet the sigmoid arteries.

2. **Sigmoid arteries.** These arteries are the several branches that enter the sigmoid mesocolon to supply the sigmoid colon.

3. **Superior rectal artery.** This artery is the most medial branch of the inferior mesenteric artery, and it enters the pelvis to supply the superior rectum. Follow it only a short distance deep to the peritoneum into the pelvis.

The **inferior mesenteric vein** is formed by tributaries that correspond to the arterial branches. The vein ascends deep to the peritoneum along the left side of the artery. It passes to the left of the duodenojejunal junction to pass deep to the tail of the pancreas, where it terminates in the splenic vein. This termination will be discussed later.

Again, note the numerous autonomic nerve fibers that distribute with branches of the superior and inferior mesenteric arteries in the layers of the peritoneum.

56. STOMACH AND SPLEEN

The **stomach** (Figure 6.11) is the part of the gastrointestinal tract that is between the esophagus and the duodenum. Often described as being shaped like the letter J, it is mainly in the upper left abdominal quadrant. However, the inferior pyloric region of the stomach crosses to the right of the midline at the level of the first lumbar vertebra. Identify the two curvatures (p. 268) of the stomach: the **lesser curvature,** which is on the right border of the stomach between the pylorus and the esophagus, and the **greater curvature,** which is on the left border. Review the peritoneal attachment of the lesser omentum between the liver and the lesser curvature. The greater curvature is much longer than the lesser curvature. Review its attachment to the greater omentum and describe the parts of the greater omentum.

The stomach is divided into the following parts: cardia, fundus, body and pylorus (p. 268). The **cardia** is the part of the stomach that

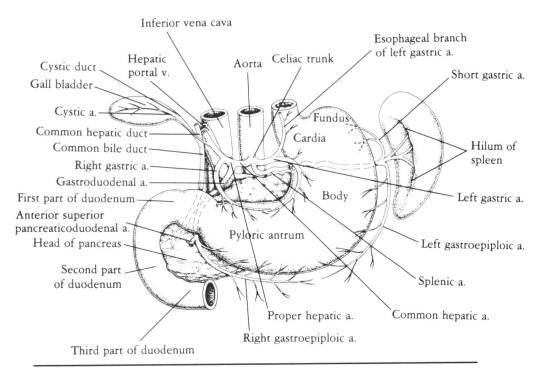

Figure 6.11

is continuous with the esophagus, and the **fundus** is the most superior part, superior to the cardia. The cardia and fundus are adjacent to the diaphragm and deep to the left costal margin. The **body** is the largest part of the stomach and is related anteriorly to the abdominal wall and posteriorly to the omental bursa. At the distal end of the body, the **angular notch** on the lesser curvature indicates the beginning of the **pylorus.** The pylorus crosses to the right of the midline with a slight upward course and is continuous with the first part of the **duodenum.** At the junction with the duodenum, the pyloric valve controls the emptying of the contents of the stomach.

The **spleen** (Figure 6.11) is not part of the gastrointestinal tract but is closely related to many of the abdominal organs and has a common blood supply (pp. 274; 289–291). It is a lymphatic organ that has a characteristic purple color. Identify the spleen posterior and to the left of the left border of the stomach. The lateral (diaphragmatic) surface of the spleen is lodged in a concavity of the diaphragm between the ninth and the eleventh ribs. The **visceral surface** is triangular and contains the **hilum** (p. 274), where the splenic vessels and autonomic nerves enter and leave the spleen. The visceral surface is related to the stomach, left kidney and left colic (splenic) flexure of the colon. The spleen is covered by the peritoneum and is attached to the stomach by the **gastrolienal ligament** and to the left kidney by the **lienorenal ligament.** When the peritoneal attachments at the hilum are grasped between two fingers, the gastrolienal ligament is ventral and the lienorenal ligament is dorsal. The lienorenal ligament contains the tail of the pancreas and the distal splenic vessels. Inferiorly, the spleen rests on the splenic flexure of the colon (phrenicocolic ligament).

57. CELIAC TRUNK

This short arterial trunk (Figure 6.11) is the most superior of the three unpaired branches of the abdominal aorta that supply the abdominal viscera (pp. 288–291). It arises from the aorta at the level of the upper border of the first lumbar vertebra and is tightly encased by the celiac ganglia and plexus of the autonomic nervous system. The courses of the three main branches of the celiac trunk are retroperitoneal until they enter a reflection of the peritoneum to reach their respective viscera. To observe the trunk, remove any remaining lesser omentum but leave the right free margin intact. Inspect the posterior wall of the omental bursa and palpate the anterior surface of the pancreas. In the midline, remove the parietal peritoneum that covers the pancreas and identify the aorta deep to the superior border of the pancreas. Then, observe the celiac trunk, which arises from the aorta at the superior border of the pancreas (Figure 6.11). The celiac trunk may be invaded by the pancreas. Clean away all peritoneum and fascia to identify the **splenic, left gastric** and **common hepatic branches** (pp. 288–289) of the **celiac trunk,** as described in the text that follows (Figure 6.11). These vessels are also covered by many autonomic nerve fibers.

A. Splenic Artery

This vessel is the largest of the three branches and follows a tortuous transverse course to the left along the superior surface of the pancreas (p. 289). It is often embedded in some of the pancreatic tissue. Distally at the tail of the pancreas, it enters the lienorenal ligament to reach the hilum of the spleen. Follow this artery along the superior surface of the pancreas, noting the numerous **pancreatic branches** that enter the gland. The two most prominent branches are the dorsal pancreatic artery (close to the midline) and the great pancreatic artery (more distal on the splenic artery). In addition, at the most distal end of the artery close to the spleen, there are short gastric arteries to the fundus and the **left gastroepiploic artery** to the greater curvature (p. 277). Identify the latter artery, which enters the greater omentum (gastrocolic ligament) at the left end of the greater curvature. In the gastrocolic ligament, it anastomoses with the right gastroepiploic artery (p. 277).

B. Left Gastric Artery

Deep to the peritoneum, observe that the left gastric artery passes posterior to the parietal peritoneum superiorly and to the left toward the cardia (pp. 288–289). At the stomach, it enters the lesser omentum and passes along the lesser curvature to anastomose with the right gastric artery, a branch of the proper hepatic artery. As the left gastric artery enters the omentum, it provides **esophageal branches** and often a **hepatic branch.** Identify these branches.

C. Common Hepatic Artery

Identify this artery posterior to the peritoneum as it crosses to the right along the superior surface of the pancreas (pp. 288–289). It is often embedded in the pancreas. When the common hepatic artery reaches the superior surface of the duodenum, it enters the lesser omentum and divides into the proper hepatic and gastroduodenal arteries (pp. 288–289).

Note that the **proper hepatic artery** ascends in the lesser omentum and it can be identified within the right free margin of the hepatoduodenal ligament on the left side of the common bile duct. Clear the course of the proper hepatic artery from the peritoneum of the hepatoduodenal ligament. The **right gastric artery** is usually the first branch of the proper hepatic artery. Identify the right gastric vessels that course in the lesser omentum along the lesser curvature. Follow the proper hepatic artery toward the hilum of the liver. It divides into the right and left hepatic arteries (pp. 288–289), which will be dissected later.

Follow the **gastroduodenal artery** as it descends posterior to the first part of the duodenum. Inferior to the duodenum, identify its division into the **right gastroepiploic** and **anterior superior pancreaticoduodenal arteries** by elevating the first part of the duodenum (pp. 290–293). Follow the right gastroepiploic artery, which enters the greater omentum (gastrocolic ligament) at the right end of the greater curvature to anastomose with the left gastroepiploic artery. Now, complete the courses of the right and left gastroepiploic arteries within the gastrocolic ligament. These vessels can be extremely small and may pass some distance from the greater curvature. Cut the gastrocolic ligament inferior to the vessels to examine the posterior wall of the lesser sac. On the posterior wall, palpate the left kidney, left suprarenal gland and pancreas.

Next, from the gastroduodenal artery, follow the anterior superior pancreaticoduodenal artery as it crosses the head of the pancreas ventrally, where it anastomoses with the anterior inferior pancreaticoduodenal artery, which arises from the superior mesenteric artery inferiorly deep to the pancreas. Other divisions of the gastroduodenal artery are the supraduodenal and retroduodenal branches to the duodenum and the **posterior superior pancreaticoduodenal artery.** Identify the latter artery, which branches from the gastroduodenal artery close to its common hepatic origin. It passes dorsal to the head of the pancreas to anastomose with the posterior inferior pancreaticoduodenal artery, which arises from the superior mesenteric artery (p. 290).

58. DUODENUM AND PANCREAS

The **duodenum** (Figure 6.12) is the first segment of the small intestine and is continuous proximally with the stomach at the pylorus and distally with the jejunum at the duodenojejunal junction (pp. 286–293). The duodenum is shaped like the letter C and is usually divided into first (superior), second (descending), third (horizontal)

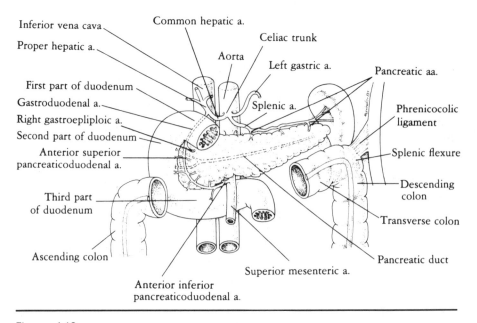

Inferior vena cava
Common hepatic a.
Proper hepatic a.
Celiac trunk
Aorta
Left gastric a.
Pancreatic aa.
First part of duodenum
Gastroduodenal a.
Splenic a.
Right gastroepliploic a.
Phrenicocolic ligament
Second part of duodenum
Anterior superior pancreaticoduodenal a.
Splenic flexure
Third part of duodenum
Descending colon
Transverse colon
Ascending colon
Pancreatic duct
Superior mesenteric a.
Anterior inferior pancreaticoduodenal a.

Figure 6.12

and fourth (ascending) parts (pp. 270–271). Note that the **first part** passes upward on the right side of the vertebral column at the level of the first lumbar vertebra and is covered by the peritoneum (hepatoduodenal ligament). Observe that the first part is related to the gallbladder and quadrate lobe of the liver superiorly. Posterior to the first part, review the courses of the **gastroduodenal artery** (and its branches), **common bile duct** and **hepatic portal vein.** The head of the pancreas is inferior. The **second part** turns inferiorly along the right side of the first, second and third lumbar vertebrae and is retroperitoneal. The transverse colon crosses the second part anteriorly and the hilum of the right kidney and the renal vessels are posterior. The head of the pancreas is to its left. At the midpoint of the second part and adjacent to the pancreas, the fused common bile duct and pancreatic duct penetrate the duodenal wall to open at the major duodenal papilla (p. 271), which can be seen by making a small vertical cut in the anterior wall. Inferiorly, note that the **third part** passes to the left across the midline at the third lumbar vertebra. It is also retroperitoneal. Posteriorly, it crosses the inferior vena cava and aorta. Anteriorly, note that the root of the mesentery and the superior mesenteric vessels cross this part. To the left of the midline, the **fourth part** ascends to the duodenojejunal flexure along the left side of the second lumbar vertebra. Its termination is covered by the peritoneum and is mobile. At its junction with the jejunum, the root of the mesentery begins and passes over the duodenum. The aorta is to the right of the fourth part.

The **pancreas** is retroperitoneal and passes horizontally across the posterior abdominal wall from the duodenum on the right to the spleen on the left (Figure 6.12). Review its position in the posterior wall of the lesser sac (omental bursa) (pp. 270–271; 289–293). The pancreas is divided into a **head, neck, body** and **tail.** Carefully remove any remaining peritoneum from its anterior surface, again noting the position of the midline celiac artery at its superior margin. Note that the **head** is the largest portion and is lodged in the C-shaped curve of the duodenum. Review the courses of the anterior and posterior arches of the pancreaticoduodenal artery on each surface of the head. The transverse colon crosses the head anteriorly. The hooklike portion of the head that projects posterior to the superior mesenteric vessels is the **uncinate process.** The **neck** is the constricted portion that overlies the origin of the **superior mesenteric artery** and the junction of the **superior mesenteric** and **splenic veins** to form the **portal vein.** Identify the origin of the superior mesenteric artery from the aorta posterior to the neck of the pancreas by elevating the pancreas (pp. 270; 277; 291). Watch for the **anterior** and **posterior inferior pancreaticoduodenal arteries,** which branch from the superior mesenteric artery (pp. 290–291). The **body** passes to the left across the aorta, left suprarenal gland and left kidney. The transverse colon and reflection of the transverse mesocolon cross the body anteriorly. Review the course of the **splenic artery** along the superior surface of the neck and body with its dorsal pancreatic and great pancreatic arteries. The inferior pancreatic artery is embedded in the inferior margin of the pancreas. The **tail** turns superiorly and enters the lienorenal ligament to reach the spleen (p. 293). In the center of the body and neck, carefully remove the tissue to identify the **pancreatic duct.** Follow this duct into the head and find its junction with the common bile duct (pp. 270–271). Elevate the pancreas and identify the **common bile duct** as it enters the gland superiorly. Again, follow these structures into the second part of the duodenum at the greater duodenal papilla. An accessory pancreatic duct is often present superior to the main duct.

59. LIVER AND GALLBLADDER

The liver is the largest gland of the body and is mainly in the upper right quadrant of the abdomen, inferior to the diaphragm and deep to the right costal margin (p. 265). The left end of the liver extends across the midline to a point just inferior to the apex of the heart.

The liver has an elaborate covering of visceral peritoneum (except for the bare area adjacent to the diaphragm and posterior wall). From the anterior body wall, identify the **falciform ligament,** which extends from the umbilicus to the visceral surface of the liver. In its inferior free border, identify the **ligamentum teres** (p. 265). Pull the diaphragmatic surface of the liver from the diaphragm and pass your hands over the liver. Follow the falciform ligament and observe the reflection of the layers of the peritoneum on the diaphragmatic surface from the body wall to form the right and left extensions of the **coronary ligament,** which extends onto the diaphragmatic surface of the right and left lobes (p. 273). Here, the **right** and **left triangular ligaments** are formed. The coronary ligament can be palpated and followed deep to the costal margin of the chest wall. The bare area of the liver is devoid of a peritoneal covering and is primarily posterior to the right triangular ligament. The medial reflection of the peritoneum at the triangular ligaments reaches the visceral surface of the liver to encircle the porta hepatis and form the lesser omentum (hepatogastric and hepatoduodenal ligaments) (p. 285). All these reflections should be understood before the liver is removed.

In preparing to remove the liver, examine the structures that pass into the liver through the right free margin of the lesser omentum: the **proper hepatic artery, portal vein** and **common bile duct** (pp. 288–289). Clean the peritoneum (hepatoduodenal ligament) from the margin of the lesser omentum to view these structures. The artery and bile duct are ventral and the portal vein is dorsal. Note that they enter or leave the liver at the porta hepatis on its visceral surface.

At the porta hepatis, identify the **right** and **left hepatic ducts,** which unite to form the **common hepatic duct** (p. 273). Inferior to this junction, observe the **cystic duct** as it joins the common hepatic duct to form the **common bile duct** in the lesser omentum. Review its distal course to the second part of the duodenum. Note that on reaching the liver, the **proper hepatic artery** divides into the **right** and **left hepatic arteries** at the porta hepatis. Identify the **cystic artery,** which supplies the gallbladder. Follow this artery to its usual origin from the right hepatic artery. Follow the **portal vein,** which distributes to the liver in a manner similar to the arteries. Review its origin deep to the neck of the pancreas and note its division into right and left branches at the porta.

At a point midway between the liver and the stomach, cut through the three structures in the free edge of the lesser omentum. Cut the artery and bile duct at a point inferior to the junction of the cystic artery and duct. Next, cut the peritoneal reflection of the liver (falciform and coronary ligaments) from the anterior body wall and diaphragm to free the liver from these attachments. Reaching over the diaphragmatic surface of the liver, cut a small segment of the inferior vena cava so that it can be removed with the liver because these structures are firmly attached to each other. The liver can now be lifted and removed. Look into the segment of the inferior vena cava that has been removed with the liver to identify the **hepatic veins.**

With the liver removed, note its wedge shape and study its diaphragmatic and visceral surfaces (p. 273). The diaphragmatic surface is related superiorly through the diaphragm to the right lung, heart and left lung. Posteriorly, this surface abuts the diaphragm and inferior vena cava and anteriorly is deep to the costal margin, xiphoid process and diaphragm. The visceral surface slopes inferior and to the left and is separated from the diaphragmatic surface by the sharp **inferior border.** Observe on the visceral surface the impression made by the stomach on the left aspect and the impression made by the right kidney on the right aspect. The lobes of the liver are easy to identify on the visceral surface. Centrally, an H-shaped area is formed by the **fissures** for the **ligamentum teres** and **ligamentum venosum** on the left and by the **fossae** for the **gallbladder** and **inferior vena cava** on the right (p. 273). Centrally between these structures, the crossbar of the H is formed by the porta hepatis. To the right of the H is the **right lobe** and to the left is the **left lobe.**

Between the ligamentum teres and the gallbladder is the **quadrate lobe** and between the inferior vena cava and the ligamentum venosum is the **caudate lobe.** Functionally, the quadrate lobe and most of the caudate lobe belong to the left lobe because they are supplied mainly by the left hepatic artery.

The **gallbladder** stores bile secretions from the liver and releases them through the cystic duct into the biliary excretory system. The gallbladder lies in the fossa on the inferior visceral surface of the liver and is covered by peritoneum reflected from the liver (pp. 272–273; 288). Identify the **fundus,** which hangs inferior to the inferior border of the liver adjacent to the first part of the duodenum, with the **body** and **neck** of the gallbladder lying in the fossa. The neck curves and continues as the **cystic duct.** Identify the junction of the cystic duct with the common hepatic duct. Note the course and origin of the **cystic artery.**

Review the entire **hepatic portal venous system,** which drains most of the gastrointestinal system, pancreas and spleen (pp. 274–277). Again, note the origin of the **portal vein** deep to the neck of the pancreas by the junction of the **superior mesenteric** and **splenic veins.** Identify the termination of the **inferior mesenteric vein** into the splenic vein. Review the course of the portal vein to the liver. Describe the alternate routes (esophageal, rectal, paraumbilical and retroperitoneal) of venous return to the right heart if portal obstruction prevents the normal flow of blood into the inferior vena cava.

60. KIDNEYS AND SUPRARENAL GLANDS

The kidneys (p. 304) are retroperitoneal and are in the paravertebral gutters opposite the twelfth thoracic and first three lumbar vertebrae (Figure 6.13). The left kidney is slightly higher than the right kidney. To view the kidneys, mobilize the ascending and descending colon to separate these parts of the colon from the fascia that covers the kidneys. Note that the kidneys and suprarenal glands are embedded in a thick, fatty layer (**perirenal fat**) that is covered externally by a membranous **renal fascia** (pp. 298–299). Carefully remove the perirenal fat and renal fascia to observe the kidneys. Identify and preserve the suprarenal glands (embedded in the perirenal fat) at the superior poles of the kidneys. Note that medially, the kidneys are in contact with the **psoas major;** posteriorly, they lie on the **quadratus lumborum** (p. 309) and **lateral** and **medial lumbocostal arches** (arcuate ligaments) (pp. 304–305). Free the kidneys from their beds to observe these posterior relationships. Review the major anterior visceral relationships of each kidney.

The lateral border of each kidney is convex and the medial border

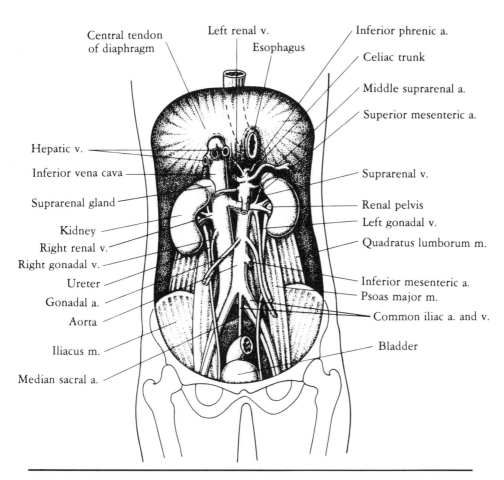

Figure 6.13

is concave, containing the **hilum.** At the hilum, identify the **renal artery, renal vein** and **ureter** (pp. 301–303). The hollow, fat-filled space that surrounds these structures at the hilum is the **renal sinus.** Follow the **renal arteries** (Figure 6.13) to the aorta and note that they arise at the level of the second lumbar vertebra. The right renal artery is longer, passing posterior to the inferior vena cava. Identify the **inferior suprarenal arteries,** which branch from the renal arteries. Next, trace the **renal veins** (Figure 6.13) to the inferior vena cava and note that they are anterior to the arteries. The left renal vein (p. 305) is longer than the right renal vein and it typically receives the left **gonadal vein** and **left suprarenal vein.** The latter two veins drain into the inferior vena cava on the right side.

The **ureter** exits the kidney at the hilum, posterior to the renal vessels (Figure 6.13). Follow the abdominal course of the ureter retroperitoneally on the psoas major until it crosses the common iliac artery to enter the pelvis (p. 305). The ureter is crossed anteriorly by the gonadal vessels on the posterior body wall.

Internal features of the kidneys can be seen better if a coronal section is made *in situ* on one side to view the organ as pages of an opened book. Note the **cortex, medulla, pyramids** and **column** (p. 300). Also, note the drainage of the **minor** and **major calices** to form the **renal pelvis,** which narrows into the **ureter.**

The **suprarenal glands** are located on the medial aspect of the superior pole of each kidney, separated by a thin layer of fat but contained in the renal fascia. Be careful to separate the suprarenal glands from the surrounding kidney, fat and fascia. Make a section through one gland and note its **cortex** and **medulla.** Blood (Figure 6.13) is supplied by the **superior suprarenal artery** (from the inferior phrenic artery), **middle suprarenal artery** (from the aorta) and **inferior suprarenal artery** (from the renal artery). Identify as many of these vessels as possible. Identify the venous drainage (a single **suprarenal vein**) to the inferior vena cava on the right side and to the left renal vein on the left side.

61. DIAPHRAGM

The diaphragm is the musculofibrous structure that separates the thorax from the abdomen. With the liver removed, the thoracic and abdominal surfaces of the diaphragm can now be studied (pp. 256–257; 306; 309). In normal respiration, the diaphragm ascends as high as the fifth rib or fifth intercostal space. Note that centrally, the diaphragm is composed of dense, fibrous tissue, the **central tendon** (Figure 6.13), which receives the insertion of the peripheral muscle fibers. The inferior vena cava passes through the caval hiatus in the central tendon at the level of the eighth thoracic vertebra. Superiorly, the fibrous pericardium is fused to the central tendon. Observe that the **muscular portion** has origins from the xiphoid process, costal cartilages of the lower six ribs and upper lumbar vertebrae. The origins from the vertebrae are represented by two crura. Follow the diaphragm posteriorly and inferiorly and identify the **right crus,** which attaches to the upper three lumbar vertebrae, and the **left crus,** which usually attaches to the first and second lumber vertebrae. Note that the right crus splits and decussates to form the **esophageal hiatus** (Figure 6.13) at the level of the tenth thoracic vertebra. Identify fibers of the **anterior** and **posterior vagal trunks** as they pass through the esophageal hiatus with the esophagus. Posterior to and between the two crura at the midline, the muscle fibers of these crura arch to form the **aortic hiatus** at the level of the twelfth thoracic vertebra. The **thoracic duct** passes through the aortic hiatus with the aorta. In addition, follow the courses of the three **thoracic splanchnic nerves** as they pass through the crura to enter the abdomen. The **caval hiatus** is located in the central tendon at the eighth thoracic vertebra.

Note that some of the diaphragmatic muscle fibers also arise from the **medial** and **lateral lumbocostal arches** (arcuate ligaments). The medial arch is a thickening of fascia superior to the psoas major between the second lumbar vertebra and the first lumbar transverse process. The lateral arch is a thickening of fascia superior to the quadratus lumborum between the first lumbar transverse process and the tip of the twelfth rib. The diaphragm is covered by the parietal pleura superiorly and the parietal peritoneum inferiorly. Study the blood and nerve (motor and sensory) supply to the diaphragm in your textbook. Also, review the movements and functions of the diaphragm in respiration.

62. POSTERIOR ABDOMINAL WALL

A. Muscles

The bony and muscular components of this wall are formed by the **five lumbar vertebrae** and the **psoas major, quadratus lumborum** and **iliacus** (p. 309). Remove any remaining parietal peritoneum from the posterior right and left walls. Be careful to protect the structures on the posterior wall deep to the peritoneum (e.g., vessels and nerves). Elevate the kidneys and note the **psoas major** on either side of the vertebral column inferior to the medial lumbocostal arch (Figure 6.13). The **psoas minor,** if present, can be identified as a long, flat tendon on the anterior surface of the psoas major. Identify the fan-shaped **iliacus,** which arises from the concavity of the iliac fossa lateral to the psoas major. The fibers of this muscle descend and blend with those of the psoas major and pass deep to the inguinal ligament to insert on the lesser trochanter of the femur. The **quadratus lumborum** (Figure 6.13) is lateral to the psoas major and extends between the twelfth rib above and the iliac crest below. The psoas major and quadratus lumborum form the paravertebral gutters that contain the kidneys.

B. Vessels

The main arterial and venous channels (Figure 6.13) on the posterior abdominal wall are the **abdominal aorta** and **inferior vena cava** (pp. 303–305; 309). The aorta enters the abdomen by passing posterior to the crura of the diaphragm at the level of the twelfth thoracic vertebra. It descends anterior to the lumbar vertebrae to the fourth lumbar vertebra, where it divides into the two **common iliac ar-**

teries. Many of the aortic branches in the abdomen have been dissected. These vessels are divided into paired parietal branches, paired visceral branches and unpaired visceral branches.

Identify examples of the paired parietal branches, represented by the inferior phrenic artery (p. 309) and four pairs of lumbar arteries on each side. The **inferior phrenic artery** is the first branch in the abdomen and it courses on the inferior surface of the diaphragm. It is small and often difficult to identify. It often gives rise to the **superior suprarenal artery.** The **lumbar arteries** (p. 309) arise at the level of the first four lumbar vertebrae. They pass deep to the psoas major and ventral to the quadratus lumborum onto the posterior body wall. Dissect into the psoas major to identify the arteries.

The paired visceral branches are the **renal, middle suprarenal** and **gonadal arteries.** These arteries have been dissected and described, except for the gonadal branches. The gonadal arteries (testicular or ovarian) are small and branch from the ventral surface of the aorta just inferior to the renal arteries (p. 305). They descend obliquely and laterally deep to the peritoneum, coursing ventrally across the ureters to enter the pelvic cavity by passing ventral to the external iliac artery. The pelvic course will be followed later.

The unpaired visceral branches are the **celiac trunk** and **superior** and **inferior mesenteric arteries.** These vessels have been dissected and should be reviewed.

The **inferior vena cava** (Figure 6.13) lies to the right of the aorta and slightly overlaps it. Observe that the vena cava is formed at the level of the fifth lumbar vertebra by the junction of the two **common iliac veins,** which drain the pelvis and lower limbs. The inferior vena cava receives the nonportal venous return from the abdomen and its main tributaries are the lumbar, right gonadal, renal, right suprarenal, inferior phrenic and hepatic veins. Locate examples of as many of these veins as possible. The hepatic veins enter as the inferior vena cava passes in the groove on the posterior surface of the liver. The vena cava passes through the central tendon of the diaphragm at the level of the eighth thoracic vertebra to enter the thorax and right atrium of the heart.

Identify, deep to the aorta, the **cisterna chyli,** which is the origin of the **thoracic lymphatic duct** (p. 306). Lumbar and intestinal channels are tributaries to the cisterna chyli.

C. Lumbar Plexus of Nerves

The **lumbar plexus** is formed by the ventral rami of the first four lumbar spinal nerves (pp. 305; 432; 435). Its cutaneous branches are distributed to the anterior and lateral walls of the lower trunk, the anterior surface of the scrotum or labia majora and the anterior and medial surfaces of the thigh. Its motor fibers are involved primarily with muscles of the lower abdominal wall and muscles in the extensor and adductor compartments of the thighs.

Cutaneous and motor branches of the lumbar plexus that distribute mainly to the trunk are the **iliohypogastric, ilioinguinal** and **genitofemoral** (its genital branch is motor to the cremaster muscle in the inguinal canal of the male) **nerves.** Identify and describe the distributions of the iliohypogastric, ilioinguinal and lateral femoral nerves deep to the peritoneum on the posterior abdominal wall. Observe their relationships to the psoas major and other muscles of the posterior wall. These nerves often fuse, making identification difficult. The iliohypogastric nerve ends anteriorly at the suprapubic area and the ilioinguinal nerve enters the inguinal canal to exit the superficial inguinal ring with the spermatic cord (or round ligament). These anterior relationships make the iliohypogastric and ilioinguinal nerves easier to identify on the anterior abdominal wall. The genitofemoral nerve is identified on the ventral surface of the psoas major.

The cutaneous and motor branches from the lumbar plexus to the lower limbs and posterior abdominal wall are the obturator, femoral and direct motor nerves to the psoas major and minor and the quadratus lumborum muscles (p. 305). Remove or reflect the psoas major laterally from the vertebral column and observe the position of the **obturator nerve** deep and medial to the muscle adjacent to the vertebral column. The obturator nerve crosses the common iliac vessel posteriorly to enter the pelvis, where it will be followed later. The obturator nerve innervates most of the adductor compartments of the thigh. The **femoral nerve** is a large nerve that can be seen leaving the inferior, lateral border of the psoas. The femoral nerve passes deep to the inguinal ligament to enter the anterior extensor compartment of the thigh, where it innervates the muscle mass of this compartment. It also has numerous cutaneous branches, primarily to the thigh. The **lateral femoral cutaneous nerve** emerges at the lower lateral aspect of the psoas major to pass deep to the inguinal ligament to enter the thigh.

D. Autonomic Nervous System

The autonomic nervous system (Fig. 6.14) in the abdomen has sympathetic and parasympathetic components (pp. 308–309).

Sympathetic Fibers. The **lumbar sympathetic trunks** can be identified as they enter the abdomen through the medial lumbocostal arches that border the vertebral column. They pass medial to the

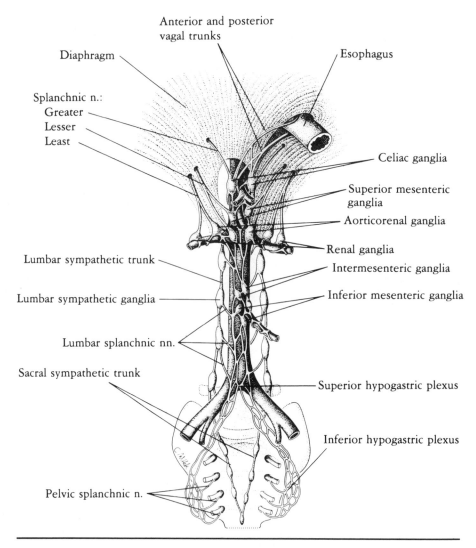

Anterior and posterior vagal trunks

Diaphragm

Esophagus

Splanchnic n.:
Greater
Lesser
Least

Celiac ganglia

Superior mesenteric ganglia

Aorticorenal ganglia

Renal ganglia

Lumbar sympathetic trunk

Intermesenteric ganglia

Lumbar sympathetic ganglia

Inferior mesenteric ganglia

Lumbar splanchnic nn.

Sacral sympathetic trunk

Superior hypogastric plexus

Inferior hypogastric plexus

Pelvic splanchnic n.

Figure 6.14

psoas major onto the ventral surface of the vertebrae. On the sympathetic trunk, identify the **lumbar sympathetic ganglia** (usually four). The first two ganglia are connected to the spinal nerves by **white rami** and **gray rami communicantes.** The lower ganglia have only gray rami. Identify the **lumbar splanchnic nerves** (usually four), which branch from the ganglia and pass medially to carry sympathetic preganglionic fibers that synapse in the collateral ganglia in the autonomic plexus on the aorta (Figure 6.14). The upper lumbar splanchnic nerves supply primarily the kidneys, gonads, descending and sigmoid colon and upper rectum through the intermesenteric and inferior mesenteric plexuses. The lower splanchnic nerves provide primarily sympathetic innervation to the pelvic viscera through the superior and inferior hypogastric (pelvic) plexuses.

Review the courses and formations of the **greater, lesser** and **least thoracic splanchnic nerves** in the thorax and follow them into the abdomen to the autonomic plexus on the aorta. These nerves provide most of the sympathetic innervation to the abdominal viscera. Dissect the greater splanchnic nerve through the crura of the diaphragm to the celiac and superior mesenteric ganglia, where the fibers synapse. The lesser and least splanchnic nerves synapse in other ganglia of the aortic plexus (superior mesenteric, renal and aortico-renal ganglia). The lumbar and thoracic splanchnic nerves carry preganglionic fibers that synapse in the collateral ganglia within the aortic plexus, which will be described. General visceral afferent fibers travel with the sympathetic motor fibers back to the spinal cord with cell bodies in the dorsal root ganglia.

Parasympathetic Fibers. Parasympathetic innervation to the viscera and glands of the abdomen to approximately the level of the left colic (splenic) flexure is provided by the vagus nerves. The vagus nerves innervate the viscera and glands supplied by the celiac, superior mesenteric, renal and gonadal arteries. From the esophageal plexus in the thorax, the vagal fibers enter the abdomen with the esophagus as **anterior** and **posterior vagal trunks** (p. 308). Identify these trunks on the surface of the esophagus (Figure 6.14) as it pierces the diaphragm to reach the cardia of the stomach. The anterior trunk follows the lesser curvature in the lesser omentum, supplying primarily the stomach. It also provides a hepatic branch to the liver and gallbladder. The posterior vagal trunk passes posterior to the cardia and provides branches to the stomach but primarily enters the celiac ganglia of the aortic plexus. Here, it distributes with sympathetic fibers in the aortic plexus to the organs supplied by the celiac, superior mesenteric, renal and gonadal arteries. These vagal trunks carry preganglionic fibers and a few visceral afferent fibers.

Parasympathetic innervation to the remainder of the abdominal viscera supplied by the inferior mesenteric artery (left colic flexure and descending and sigmoid colon) is provided by the **pelvic splanchnic nerves.** (Note that the use of the word splanchnic here does not describe sympathetic innervation). These nerves are in the pelvis and will be dissected later. They are preganglionic fibers derived from the second, third and fourth sacral levels of the spinal cord. The pelvic splanchnic fibers supply the pelvic viscera and ascend deep to the peritoneum into the abdomen over the sacral promontory to reach the terminal parts of the large colon, which is supplied by the inferior mesenteric artery. Note that all the para-sympathetic nerves in the abdomen (vagal or pelvic splanchnic nerves) have most of their postganglionic cell bodies in the visceral wall. Thus, the postganglionic fibers are extremely short.

Aortic Autonomic Plexus. The aortic autonomic plexus (Figure 6.14) receives and distributes most of the fibers of the sympathetic and parasympathetic nerves described previously. The plexus is located on the surface of the abdominal aorta. In the plexus are collateral (prevertebral) sympathetic ganglia (p. 308), where pre-ganglionic sympathetic fibers in the thoracic and lumbar splanchnic nerves synapse with the postganglionic fibers (except for the sympathetic fibers to the adrenal medulla). The parasympathetic pregan-glionic fibers (vagal and pelvic splanchnic nerve fibers) enter and pass through the plexus without synapsing until they reach the terminal ganglia on or in the viscera. From the aortic plexus, the sympathetic (postganglionic) and parasympathetic (still pregangli-onic) fibers distribute within many of the peritoneal layers to the organs via the paired and unpaired vascular branches of the aorta. Examples of these perivascular fibers were observed on the surfaces of many arteries as the peritoneum was removed from them. Visceral afferent fibers are numerous in all aspects of the aortic plexus but have their cell bodies in sensory dorsal root ganglia.

The **aortic plexus** is a continuous and extensive system of sympa-thetic ganglia and sympathetic and parasympathetic fibers that stream up and down the aorta. The aortic plexus receives fibers from the thoracic splanchnic (greater, lesser and least) nerves (sympa-thetic), lumbar splanchnic nerves (sympathetic) and vagus nerves (parasympathetic). The plexus is divided into regions (which are not totally discernible) based on the increased density of nerves at the origins of the paired and unpaired visceral branches of the aorta. Thus, celiac, superior mesenteric, aorticorenal, renal, intermesen-teric, gonadal and inferior mesenteric plexuses and ganglia are pres-ent. Observe the one or two **celiac ganglia** and plexus at the base of the celiac artery. These ganglia are the largest and receive mainly the fibers of the posterior **vagal trunk** and **greater thoracic splanchnic nerves.** The fibers of the greater splanchnic nerves synapse in the celiac ganglia and also send fibers to the superior mesenteric ganglia, where they synapse. The **fibers of the lesser** and **least thoracic splanchnic nerves** synapse in the renal, aortico-renal and gonadal plexuses. The **fibers of the upper lumbar splanchnic nerves** synapse in the intermesenteric and inferior mesenteric plexuses, but the **fibers of the lower lumbar splanch-nic nerves** synapse in the superior and inferior hypogastric (pelvic) plexuses in the pelvis. The superior hypogastric plexus passes over the sacral promontory and is the connection of autonomic fibers that pass between the abdomen and the pelvis via the hypogastric nerves. The vagal fibers enter the aortic plexus at the celiac ganglion without synapsing and descend through the aortic plexus inferiorly as far as the renal and gonadal plexuses. The vagal fibers synapse in terminal ganglia on or in the viscera. Remember that inferior to the splenic flexure of the colon, the pelvic splanchnic nerves provide the parasympathetic supply to the descending and sigmoid colon and pelvic viscera.

CHAPTER SEVEN
PERINEUM

63. BOUNDARIES AND LANDMARKS

The perineum is the inferior outlet of the pelvis and is separated from the pelvic cavity by the muscular pelvic diaphragm, which forms the pelvic floor. Parts of the gastrointestinal, genital and urinary systems pass through the pelvic floor to reach the perineum. The perineum is shaped like a diamond and bounded by the **symphysis pubis** anteriorly, **ischiopubic rami** anterolaterally, **sacrotuberous ligaments** posterolaterally and **coccyx** posteriorly (p. 399). The perineum is divided into an anterior urogenital triangle and a posterior anal triangle by an imaginary line that extends between the two ischial tuberosities. Before beginning the study of these regions, review the following bony and ligamentous landmarks: the **coccyx, sacrum, ischial tuberosities** and **spine, ischiopubic rami, symphysis pubis** and **sacrospinous** and **sacrotuberous ligaments** (pp. 397–402). It is important in these and subsequent dissections of the pelvis to review and compare the differences between the two sexes.

64. ANAL TRIANGLE

Place the body in the prone position and abduct the lower limbs. Use Figure 7.1 to make the following skin incisions: first, make a circular cut around the anal opening. Then, make two straight cuts, the first cut posteriorly to the coccyx and the second cut anteriorly to a point

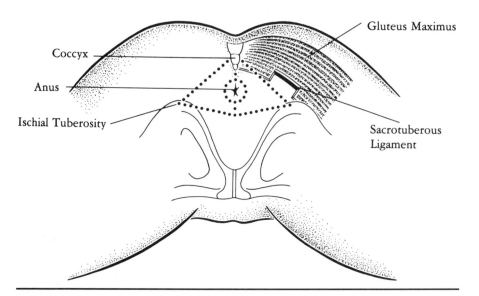

Figure 7.1

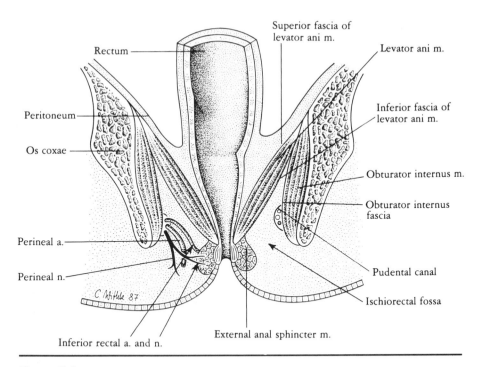

Figure 7.2

in the midline between the ischial tuberosities. From the latter point, make two lateral cuts out to the ischial tuberosities. Make the final cuts along a line between the tuberosity and the coccyx on each side of the body.

In the anal triangle, remove the skin and dense superficial fascia from the area on either side of the anal canal, noting the large fatty deposits around the anal region. Most of the anal triangle is formed by two large **ischiorectal fossae** (Figure 7.2) on each side of the rectum and anal canal (pp. 323; 338). These fossae are shaped like a wedge and form deep spaces filled with fatty deposits. Remove the fat from the fossae, being careful to preserve the **inferior rectal vessels and nerves** (Figure 7.2), which pass medially through the center of each fossa to the lower digestive tract (p. 323). Use the probe or fingers to locate these vessels and nerves before removing the fat. After the fat has been removed, examine the boundaries of the ischiorectal fossa (Figure 7.2). The **base** is formed by the overlying skin and connective tissue. The **medial wall** is formed by the sloping inferior surface of the **levator ani** and its inferior fascia and the **external anal sphincter** (pp. 322; 336–338). Anteriorly, the fossa meets the posterior free border of the urogenital diaphragm (in the urogenital triangle) and continues as a small, anterior recess superior to the diaphragm. Palpate this deep recess with one or two fingers. **Posteriorly,** the fossa is bounded by the rounded margin of the **gluteus maximus** and the deeper **sacrotuberous ligament.** If the gluteal region has not been dissected, palpate this ligament by placing your fingers around the posterior edge of the gluteus maximus until you feel the dense ligament. Project to the surface of the body the attachments of this ligament. Then, expose this ligament and the pudendal canal (discussed later) by cutting away a 5-cm segment of the gluteus maximus from the sacrotuberous ligament close to the ischial tuberosity (pp. 324–325; 338) (Figure 7.1).

The **lateral wall** is formed by the **obturator internus** muscle and fascia (Figure 7.2). Observe this muscle and fascia on the lateral wall of the ischiorectal fossa. Note that a splitting of obturator fascia forms the **pudendal canal** (Figure 7.2), which transmits the internal pudendal vessels and pudendal nerve branches from the pelvis into the perineum (pp. 323–325; 337–339). Identify the pudendal canal by tracing the inferior rectal vessels and nerves to the lateral wall, where they course in the pudendal canal deep to the sacrotuberous ligament. Slit open the canal to identify the vessels and nerves. The pudendal nerve (S2 to S4) is derived from the sacral plexus and is the primary motor and sensory innervation for the urogenital and anal triangles of the perineum. The pudendal nerve divides in the canal into the **perineal nerve, dorsal nerve of the penis** (or **clitoris**) and **inferior rectal nerves** (Figure 7.2). Identify and follow these nerves into the anal triangle. Note the following features: the inferior rectal nerves traverse medially in the ischiorectal fossa to innervate the external anal sphincter (pp. 324; 337); the perineal nerve enters the urogenital triangle by passing superficial to the urogenital diaphragm (pp. 324; 337); and the dorsal nerve enters the urogenital triangle by piercing the posterior border of the urogenital diaphragm (pp. 324–325; 327). The latter nerve maintains its distal course along the lateral wall of the fossa, embedded in the obturator fascia to reach the urogenital diaphragm. The first two nerves course through the fascia within the ischiorectal fossa. The functions of the perineal and dorsal nerves will be described in connection with the urogenital triangle. Identify the branches of the **internal pudendal vessels** that are similarly named (perineal, inferior rectal and dorsal artery of the penis or clitoris) and are distributed with their associated nerves.

In the midline of the anal triangle, clean and identify the **external anal sphincter** (p. 322), which extends posteriorly from the coccyx to pass around the anus to attach to the perineal body anteriorly. This sphincter has a superficial, fusiform part and a deeper, circular part. The external anal sphincter is supplied by the inferior rectal vessels and nerves. The deeper position of the pelvic diaphragm was previously noted and this muscular diaphragm will be dissected with the pelvis.

65. UROGENITAL TRIANGLE

The **urogenital triangle** is the anterior division of the perineum and is bounded anteriorly and laterally by the symphysis pubis and is-chiopubic rami, respectively. Passing through this area are the terminal portions of the genital and urinary systems. The urogenital triangle classically has been divided into superficial and deep perineal spaces bounded by fascial and muscular components of the triangle. In the laboratory, these boundaries (described later) are not always clearly demarcated and it is important to concentrate your studies on the contents and relationships within these two spaces of the urogenital triangle. The superficial boundary of the superficial space is the membranous layer of the superficial fascia (superficial perineal fascia). This fascial layer is continuous with the deep membranous layer of the superficial fascia identified previously in the lower abdominal wall. In the perineum, the membranous fascia has strong attachments to the inguinal ligament, ischiopubic rami and posterior border of the urogenital diaphragm of the urogenital triangle. Review these attachments in your textbook. Review the communication between the superficial space and the anterior abdominal wall deep to the membranous fascia on either side of the symphysis pubis. Understand the relationship of this fascia to the scrotum or labia majora. The deep boundary of the superficial space is the **perineal membrane** (inferior fascia of the urogenital diaphragm) (pp. 322; 336). The deep perineal space is a muscular layer composed of the urogenital diaphragm, which is bounded by the perineal membrane and the superior fascia of the urogenital diaphragm. The contents of the urogenital triangle will be considered separately for the male and female.

A. Male

Complete the skin incisions shown in Figure 7.3 and reflect the skin laterally from the base of the penis and the scrotum. Do not disturb the superficial fascia deep to the skin at this time. In the male, the membranous layer within the urogenital triangle becomes the tunica dartos layer of the scrotum and the superficial subcutaneous tissue of the penis. The membranous layer forms the superficial boundary of the superficial space with firm attachments laterally and posteriorly

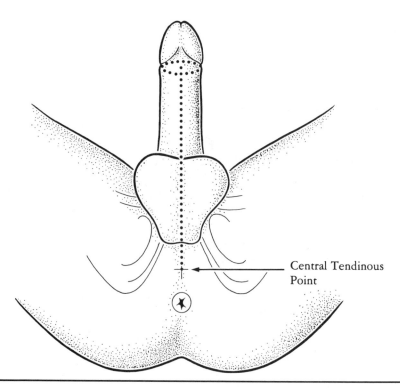

Central Tendinous
Point

Figure 7.3

to the ischiopubic rami and perineal membrane of the urogenital diaphragm, respectively, as described previously.

External Genitalia. The **external genitalia** of the male are composed of the **body,** or **shaft,** of the **penis** and the **scrotum.** Make a shallow midline incision along the ventral surface of the penis and reflect the skin around the shaft (Figure 7.3). If present, note the **prepuce** that covers the **glans penis** distally (pp. 320–321). Deep to the skin is the **subcutaneous connective tissue layer.** In this layer, watch for the single **superficial dorsal vein** (pp. 320–321) at the midline on the dorsal surface of the penis. This vein drains proximally into the external pudendal veins of the lower abdominal wall and thighs. Deep to the subcutaneous connective tissue layer is the **deep fascia** (Figure 7.4), which encircles the cavernous bodies and continues proximally into the superficial space as a muscular fascia. In the dorsal midline deep to the deep fascia, identify the **dorsal arteries** and **nerves** of the penis on each side of the midline and the single **deep dorsal vein** in the midline (pp. 320–321). This vein drains deeply into the prostatic venous plexus in the pelvic cavity by passing between the symphysis pubis and the urogenital diaphragm.

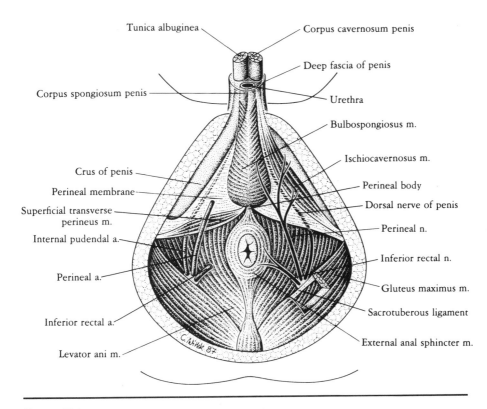

Tunica albuginea

Corpus cavernosum penis

Deep fascia of penis

Corpus spongiosum penis

Urethra

Bulbospongiosus m.

Ischiocavernosus m.

Crus of penis

Perineal body

Perineal membrane

Dorsal nerve of penis

Superficial transverse perineus m.

Perineal n.

Internal pudendal a.

Inferior rectal n.

Perineal a.

Gluteus maximus m.

Sacrotuberous ligament

Inferior rectal a.

External anal sphincter m.

Levator ani m.

Figure 7.4

Follow the dorsal vessels and nerves proximally to the symphysis pubis.

The body of the penis is formed by three cylindrical, cavernous (Figure 7.4) bodies that engorge with blood during erection. Note that they are tightly enclosed by a dense, fibrous layer (tunica albuginea). Two of these masses (**corpora cavernosa penis**) can be identified on the dorsum of the penis and one is ventral (**corpus spongiosum**) (pp. 310–311; 313–314; 322). The corpus spongiosum transmits the **penile urethra** within its center and expands distally, capping the blunt ends of the corpora cavernosa as the **glans penis** (p. 311). Blunt dissect and cut into the fibrous layer that binds together these three bodies, separating them from each other.

Superficial Perineal Space. The boundaries of this space have been described. Note that the membranous layer has been removed. Carefully tease through and remove the fatty tissue that fills the superficial space. Follow the **anterior scrotal nerves** that enter the scrotum from the anterior abdominal wall via the inguinal canal. These nerves are branches of the ilioinguinal nerve (lumbar plexus) (pp. 320–321). Posteriorly, trace the branches of the **perineal nerve**

(a branch of the pudendal nerve) (Figure 7.4), which pass superficial to the urogenital diaphragm into the superficial space from the anal triangle (pp. 323–325). In the superficial space, the perineal nerve branches into **posterior scrotal nerves** to the scrotal sac, muscular branches to the five muscles of the urogenital triangle (discussed later) and the nerve to the bulb of the penis. The latter two groups of nerves are difficult to identify in the dissection.

Next, you will follow the corpora cavernosa and corpus spongiosum deeply into the superficial perineal space to dissect the root of the penis. Separate completely the two halves of the scrotum by cutting and separating the scrotal septum. Dissect the course of the **corpora cavernosa** along the ischiopubic rami on each side and the **corpus spongiosum** into the superficial space as described in the following paragraphs.

The roots of the two corpora cavernosa originate as the **crura of the penis** (pp. 322–323). Identify the crura (Figure 7.4) on the lateral aspects of the triangle along the ischiopubic rami. Identify on the surface of each crus of the penis the **ischiocavernosus** muscles and note that their fibers extend a short distance onto the shaft of the penis, where they insert into each corpus cavernosum (Figure 7.4).

Follow the corpus spongiosum from the shaft of the penis into the superficial space. Observe that in the midline, it enlarges to form the **bulb of the penis** (pp. 322–323). The bulb is also composed of cavernous tissue and transmits the urethra from the deep perineal space into the corpus spongiosum. Identify the **bulbospongiosus** overlying the bulb. This muscle arises from the perineal body (central tendinous point) posteriorly, with fibers passing forward over the surface of the bulb (Figure 7.4). Some of the fibers insert into the perineal membrane and others extend onto the shaft of the penis, where they encircle the corpus spongiosum and corpora cavernosa to insert on the dorsum of the penis. Reflect a small area of the muscle to view the bulb of the penis.

A third pair of muscles in the superficial space, often absent or difficult to identify, is the superficial transverse perineus (pp. 322; 325). This pair of muscles attaches to the perineal body centrally and passes laterally to the ischial tuberosities. It is not necessary to identify these muscles.

All the muscles described in the superficial space (ischiocavernosus, bulbospongiosus and superficial transverse perineus) are innervated by the muscular branches of the perineal nerve. Identify the **central tendinous point** (perineal body), which is at the midpoint of the perineum between the anus and the scrotal sac (Figure 7.4). It is a dense, supportive, connective tissue in which the muscular fibers of the external anal sphincter, superficial transverse perineus and bulbospongiosus interdigitate.

On either side of and deep to the bulb of the penis, identify the dense **perineal membrane** (inferior fascia of the urogenital diaphragm) (p. 322). With a probe or fingers, note the strength provided by this membrane as it spans across the ischiopubic rami (Figure 7.4). The perineal membrane covers the inferior surface of the urogenital diaphragm and forms the deep boundary of the superficial perineal space.

Deep Perineal Space. It is important to note that the descriptions that will be given for the structures in the deep perineal space are, for the most part, not applicable in the cadaver because your work does not include removal of structures within the superficial space. Therefore, a full understanding of these relationships from your textbook and atlas is essential. The deep space is formed primarily by the muscular **urogenital diaphragm.** This diaphragm is divided into two muscles, the sphincter urethrae anteriorly and deep transverse perineus posteriorly (pp. 324–325). The muscular branches of the perineal nerve innervate the two muscular parts of the diaphragm, which with its fascia forms a strong support for the pelvic viscera. Understand that the diaphragm is almost horizontal in the anatomic position. Palpate the urogenital diaphragm by placing your thumb on the perineal membrane in the superficial space and one or two fingers in the anterior recess of the ischiorectal fossa within the anal triangle. Note the strength of the diaphragm between the fingers. The fibers of the deep transverse perineus comprise the posterior free edge of the diaphragm between the ischial tuberosities. More anteriorly, these fibers form the sphincter urethrae. The membranous urethra passes through this anterior portion of the diaphragm to enter the bulb of the penis, which is attached to the inferior surface of the diaphragm. At the point where the urethra penetrates the diaphragm, a circular investment of fibers forms a sphincter around the urethra just inferior to the prostate gland, which lies superiorly in the pelvis.

Inferiorly, the urogenital diaphragm is covered by its dense fascia, the **perineal membrane.** This structure was identified in the floor of the superficial space. The perineal membrane is continuous around the posterior and anterior borders of the diaphragm with the superior fascia of the urogenital diaphragm. Anteriorly, this continuation thickens to form the transverse perineal ligament. It is separated by a small space from the symphysis pubis, which is covered by the arcuate pubic ligament. The **deep dorsal vein** of the penis passes through this space to terminate in the prostatic venous plexus in the pelvis. The two bulbourethral glands are embedded in the sphincter urethrae on either side of the membranous urethra. The ducts from the gland penetrate the perineal membrane to empty into the urethra as it passes through the bulb of the penis.

Also passing in the deep space through the lateral aspects of the diaphragm along the ischiopubic rami are the continuations of the internal pudendal vessels and the dorsal nerve of the penis (pp. 324–325). Because they are embedded in muscle, these vessels and nerve will not be dissected in the deep space. The dorsal nerve of the penis (branch of the pudendal nerve) enters the deep perineal space from the ischiorectal fossa by penetrating the posterior free edge of the urogenital diaphragm (Figure 7.4). In the lateral wall of the ischiorectal fossa, review and identify the course of the dorsal nerve of the penis as the most lateral branch of the pudendal nerve lying against the obturator internus.

The terminal courses of the internal pudendal vessels parallel the dorsal nerve in the deep space. Branches of the artery in the deep space include the artery to the bulb, urethral artery, deep artery of the penis to the corpora cavernosa and dorsal artery of the penis. The **dorsal artery** and **dorsal nerve of the penis** leave the deep space anteriorly, where they can be identified on the dorsum of the penis. They follow the dorsum to the glans penis, supplying the superficial and deep aspects of the penile body. Review the distributions of these structures.

B. Female

Complete the skin incisions shown in Figure 7.5, reflecting the skin laterally from the labia majora. Do not dissect deeply into the superficial fascia at this time. In contrast to the male, note the increased fatty content of the female perineum, especially in the labia majora (there is no fatty component in the wall of the scrotum). The membranous layer of the superficial fascia enters the urogenital triangle from the lower abdominal wall. It terminates by attaching to the skin of the labia majora and to the inguinal ligament, ischiopubic rami and urogenital diaphragm, as in the male. The membranous fascia also forms the superficial boundary of the superficial perineal space in the female.

External Genitalia. In the female, there is no fusion of the genital swellings; therefore, a midline cleft persists. This midline separation is the **vestibule** and through it the **urethra, vagina** and ducts of the greater vestibular gland open to the exterior (pp. 334–335). Later-

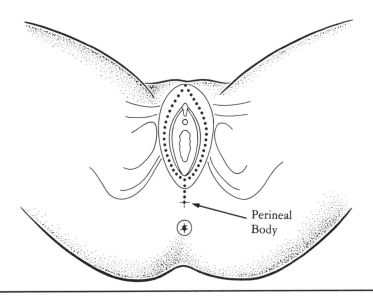

Figure 7.5

ally, the vestibule is bounded by the **labia majora** and **minora.** The **labia majora** are the rounded lateral folds of skin that contain large amounts of fat. They receive cutaneous innervation from the anterior and posterior labial nerves (described later). Anteriorly, the labia majora blend with each other to form the **mons pubis,** the rounded, fatty bulge above the pubis. The labia meet across the midline superior to the clitoris as the anterior labial commissure. There is a less definite posterior labial commissure where the labia majora meet posteriorly between the vestibule and the anus (p. 334). The inner surfaces of the labia are smooth and hairless. Adjacent to them are two smaller cutaneous folds, the **labia minora** (p. 334). These labia are devoid of fat and are thin. The labia minora divide dorsal to the clitoris to form the **prepuce of the clitoris** and ventral to the clitoris to form the **frenulum** (pp. 334–335).

The clitoris is similar in structure to the penis but is much smaller. Thus, its components are not as easily dissected. Most of the body of the clitoris is formed by the erectile bodies of the two corpora cavernosa (p. 335). There is no corpus spongiosum in the female because the urethra courses separately from the clitoris. The small glans clitoridis that caps the corpora cavernosa is formed by the anterior ends of the cavernous bulb of the vestibule (pp. 334–335) (discussed later). The dorsal nerves and arteries on the clitoris are much smaller than their counterparts in the male and are therefore more difficult to dissect. The **external urethral orifice** (p. 334) opens into the vestibule between the clitoris and the vagina (Figure 7.6).

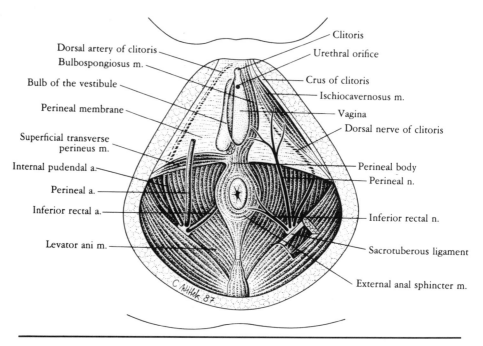

Dorsal artery of clitoris
Bulbospongiosus m.
Bulb of the vestibule
Perineal membrane
Superficial transverse perineus m.
Internal pudendal a.
Perineal a.
Inferior rectal a.
Levator ani m.

Clitoris
Urethral orifice
Crus of clitoris
Ischiocavernosus m.
Vagina
Dorsal nerve of clitoris
Perineal body
Perineal n.
Inferior rectal n.
Sacrotuberous ligament
External anal sphincter m.

C. Mittek 87

Figure 7.6

Superficial Perineal Space. Complete the removal of the superficial fatty tissue within the superficial space. Identify the branches of the **anterior labial nerves** (branches of the ilioinguinal nerve), which enter the fascia of the superficial space anteriorly from the anterior abdominal wall. In the fascia, identify the **perineal nerve** (Figure 7.6), which enters the superficial space from the ischiorectal fossa by passing superficial to the urogenital diaphragm (pp. 336–337). In the superficial space, the perineal nerve forms **posterior labial nerves** to the labia and muscular branches to the muscles of the triangle. Identify the labial nerves within the fascia.

Identify the **body of the clitoris** by removing the fascia from its lateral surface and root at the symphysis pubis (p. 337). Identify the two **corpora cavernosa** that form the body of the clitoris. Along the ischiopubic rami, follow the corpora cavernosa deeply into the superficial perineal space as the **crura of the clitoris** (Figure 7.6). Each crus is covered by the **ischiocavernosus** muscles, which are similar to those of the male. Because there is no midline fusion, the **bulb of the vestibule** (p. 337) is divided into two bodies (Figure 7.6). They are just lateral and deep to the labia majora and vagina and are covered by the **bulbospongiosus** laterally. Remove the fascia that covers the lateral surface of the labia majora to identify this muscle.

Make a small cut in the bulbospongiosus and observe deeply the bulb of the vestibule. The bulb is composed of erectile tissue and is shaped like a half-pear. A small extension of erectile tissue extends superiorly along the ventral surface of the clitoris, forming the glans clitoridis. At the posterior ends of the bulb, the greater vestibular glands (p. 336) are lodged in the superficial space. The bulbospongiosus (Figure 7.6) covers the bulb and has its origin from the perineal body, with its insertions on the perineal membrane, corpora cavernosa and dorsum of the clitoris. The superficial transverse perineus (often absent) attaches between the perineal body and the ischial tuberosities (p. 337).

The **perineal body** (Figure 7.6) in the female is located centrally between the anus and the vagina (p. 337). It is a strong connective tissue interdigitation of the bulbospongiosus, external anal sphincter and superficial transverse perineus. It is an important support for the vagina and pelvic viscera.

The **perineal membrane** can be identified with the probe or fingers as a strong connective tissue layer between the bulb of the vestibule and the ischiopubic rami (Figure 7.6). The perineal membrane spans across the ischiopubic rami to form the inferior fascia of the urogenital diaphragm. It also is a strong support for the pelvic viscera in the female.

Deep Perineal Space. As in the male, the structures in the deep space are not easily dissected and a thorough understanding should be obtained from your textbook and atlas (p. 338). The deep space is occupied primarily by the muscular urogenital diaphragm and its fascia, as in the male. Palpate this diaphragm from the ischiorectal fossa, as described in the male. It is divided into the deep transversus perineus and sphincter urethrae. Because the vagina and urethra pass through the urogenital diaphragm in the female, the sphincter urethrae is more fragmented and does not form a total sling between the ischiopubic rami, as it does in the male. Some of the fibers of the diaphragm blend with and support the walls of the vagina. These two muscles of the urogenital diaphragm are innervated by branches of the perineal nerve. The diaphragm is enclosed by the perineal membrane inferiorly and the superior fascia of the urogenital diaphragm superiorly. These two layers are continuous at the anterior and posterior borders of the muscle, forming the transverse perineal ligament anteriorly. The perineal membrane is pierced by the vagina and urethra.

In the lateral aspects of the urogenital diaphragm, but not easily dissected, the dorsal nerve of the clitoris and the terminal parts of the internal pudendal vessels (p. 338) course in the fibers of the urogenital diaphragm along the ischiopubic rami (Figure 7.6). The branches from the artery in the deep space include the artery of the bulb of the vestibule, urethral artery and deep and dorsal arteries of the clitoris. The dorsal artery and nerve pass onto the dorsum of the clitoris. Review the relationships and structures of the deep space.

CHAPTER EIGHT
PELVIS

66. BOUNDARIES AND LANDMARKS

The pelvic cavity, or basin, is continuous with the inferior aspect of the abdominal cavity and is usually divided into two parts: the major pelvis and the minor pelvis. The major (false) pelvis, the more superior part, is continuous with the abdominal cavity and contains coils of intestines. It is bounded laterally by the **alae of the iliac bones.** The major pelvis narrows inferiorly and is continuous with the minor (true) pelvis at the **superior pelvic aperture** (inlet), which is bounded by the **terminal line** of the pelvis. Identify the **terminal line** on the skeleton, formed on each side by the **pecten of the pubis** (iliopectineal line), **arcuate line of the ilium** and **sacral promontory** (pp. 399–400; 402). The lowest aspect of the pelvis is separated from the perineum by the muscular pelvic diaphragm, which fills the inferior pelvic outlet. Note that the inferior outlet is almost horizontal, whereas the superior inlet is turned obliquely forward and superiorly to provide support for the abdominal viscera.

67. BONY PELVIS

Before beginning the dissection of the soft structures of the pelvis, review the bones that form the **os coxae** (p. 400). The two os coxae articulate anteriorly with each other at the **symphysis pubis** and posteriorly with the **sacrum** at the sacroiliac joints, thus forming a complete circle of bone. Identify the **ilium, pubis** and **ischium** and observe the following landmarks on the medial surfaces of these bones (pp. 397; 399–402):

    1. Ilium. **Ala** and **crest, fossa, anterior** and **posterior superior spines, anterior** and **posterior inferior spines, arcuate line** and the roughened **tuberosity** for articulation with the sacrum.

    2. Ischium. **Inferior ramus, spine, greater** and **lesser sciatic notches; tuberosity** and **obturator foramen.**

    3. Pubis. **Superior** and **inferior rami** (the inferior pubic ramus is continuous with the ischial ramus to form the **ischiopubic ramus**), **crest, tubercle, obturator foramen** and **groove** and **pecten** (pectineal line).

68. PERITONEUM

The parietal peritoneum that lines the abdominal wall continues into the pelvic cavity. However, it does not reach the pelvic diaphragm (floor). Thus, many of the viscera are inferior to and are only partially covered by the peritoneum, being embedded in a dense **en-**

dopelvic fascia (extraperitoneal connective tissue). Many of the blood vessels and nerves run through the endopelvic fascia to reach the viscera. The transversalis fascia also continues into the pelvis from the abdominal wall.

First, examine the position of the peritoneum and its relationships to the viscera. In the male pelvis (pp. 310–311) with the cadaver in the supine position, follow the peritoneum from the anterior abdominal wall into the pelvic cavity onto the superior surface of the **bladder** posterior to the symphysis pubis. Follow the peritoneum to the fundus at the posterior curvature of the bladder, where the peritoneum sweeps onto the anterior surface of the **middle third of the rectum.** Follow the sharp **rectovesical** (sacrogenital) **peritoneal fold,** which passes from the posterior surface of the bladder to the sacrum on either side of the rectum. Laterally, between this fold and the rectum, a **pararectal fossa** is formed on each side of the rectum. These two fossae are continuous anterior to the rectum at the **rectovesical pouch,** separating the rectum from the bladder. A fold of peritoneum, formed by the ductus deferens as it crosses the rectovesical fold, can often be seen close to the bladder.

In the female pelvis (pp. 327; 329) with the cadaver in the supine position, note that the course of the peritoneum into the pelvic cavity across the pelvic viscera is more complex than in the male. This greater complexity results from the positioning of the female reproductive organs between the bladder and the rectum. Observe the positions of the **uterus, uterine tubes** and **ovaries.** All these structures are covered by a fold of the peritoneum that is elevated from the pelvic floor. This fold of peritoneum is called the **broad ligament.** Follow the peritoneum over the superior surface of the bladder to its posterior border. At this point, the peritoneum reflects onto the inferior, vesical surface of the uterus and continues over the fundus onto the superior, intestinal surface of the uterus. Note that the body and fundus of the uterus bend anteriorly over the bladder. From the intestinal surface of the uterus, the peritoneum reaches the **posterior fornix** of the vagina to sweep onto the **middle third of the rectum.** The position of the uterus between the bladder and rectum produces two peritoneal pouches: the **vesicouterine** (anterior) and **rectouterine** (posterior) pouches.

At the lateral margins of the uterus, the double-layered fold of the peritoneum that projects upward from the pelvic floor forms the **broad ligament.** Identify its subdivisions (pp. 330–331), which are named according to the parts of the reproductive system that they cover: the **mesovarium,** which surrounds the ovary; the **mesosalpinx,** which surrounds the uterine tube; and the **mesometrium,** which is attached to the lateral sides of the uterus. Follow the **round**

ligament as it passes deep to the anterior layer of the mesometrium to the deep inguinal ring. Identify the **ovarian ligament** and describe its course in the broad ligament. The uterine vessels and ureter course in the endopelvic fascia at the base of the broad ligament. The **suspensory ligament of the ovary** can be seen passing over the pelvic brim and attaching to the superior pole of the ovary. The ovarian vessels, nerves and lymphatic vessels are in this ligament. These vessels and nerves will be described in a later dissection.

69. RECTUM

The rectum and anal canal (pp. 311; 329) are the terminal parts of the gastrointestinal tract. The rectum enters the pelvis from its continuity with the sigmoid colon at the pelvic brim. The upper two-thirds of the rectum are partially covered by the peritoneum, whereas the lower third is completely devoid of a peritoneal covering. This lower third is embedded in a layer of rectal fusion fascia that encloses the rectal blood vessels (discussed later). Observe that the rectum takes a tortuous course in the pelvis, bending to the right and then to the left. **Transverse rectal folds** are prominent as the rectum courses from side to side. In the male, the **prostate gland** and **bladder** are covered by the peritoneum anterior to the rectovesical pouch and in the female the **vagina** is anterior to the lower third of the rectum. The blood supply to the rectum will be described later.

The **anal canal** is distal to the rectum and begins as the rectum turns inferiorly and posteriorly to pass through the genital hiatus of the levator ani portion of the pelvic diaphragm. At the pelvic diaphragm, the rectum and anal canal join at right angles. The anal canal is not visible in the pelvis, being located inferior to the pelvic diaphragm in the anal triangle. Anteriorly, the anal canal is supported by the central tendinous point (perineal body). Review in your textbook the mucosal features of the anal canal, noting the rectal venous plexus and its relationship to internal and external hemorrhoids. These features (anal columns and anal valves) of the anal canal can be observed in the midsagittal section of the pelvis (described later). Review the role of the venous plexus in alternate portal venous return to the heart when the hepatic portal venous return is obstructed by liver disease.

70. MIDSAGITTAL SECTION

To facilitate further dissection, it is necessary to make a midsagittal section of the pelvis. This cut will remove the right half of the pelvis, right lower posterior abdominal wall and attached lower limb. Use Figure 8.1 to make the following cuts: first, abduct the lower limbs.

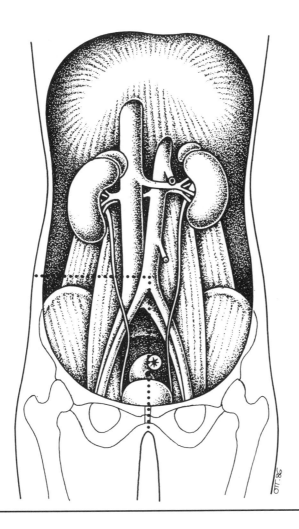

Figure 8.1

Use the scalpel to cut the soft structures of the perineum and pelvis and use a saw to cut through the bony parts of the pelvic girdle and lumbar vertebrae in the midsagittal plane. In the male, cut and separate the corpora cavernosa of the penis and use the scalpel to make a median cut through the corpus spongiosum and urethra. In the female, make a cut through the median plane of the clitoris and vestibule. Next, in both sexes, begin your saw cut at the symphysis pubis. Elevate the body and continue the cut superiorly in the midline through the coccyx and sacrum. Use the scalpel to make midline cuts in the pelvic viscera. Pull the rectum to the left as much as possible to keep most of it away from the saw cut. Carry the midline cut superiorly to the fourth lumbar vertebra. From this point, make a lateral, transverse cut across the right posterior abdominal body wall superior to the right iliac crest. This cut will necessitate severing the muscles, vessels, nerves and ureter that pass across the posterior abdominal wall. Most of the transverse cut can be made with the scalpel. Mobilize the right kidney and any parts of the large colon and move them away from the transverse cut.

71. URETER AND URINARY BLADDER

The abdominal course of the ureter (p. 305) was dissected on the posterior abdominal wall. Review this abdominal course deep to the peritoneum on the surface of the psoas major. Note that the ureter crosses the pelvic brim at the bifurcation of the common iliac artery as it enters the pelvis.

In the male, remove the peritoneum that covers the **ureter** and follow it retroperitoneally on the lateral pelvic wall. It passes the medial side of the obturator nerve to reach the posterolateral surface of the bladder (pp. 310; 316). At the bladder, the ureter crosses the superior surface of the seminal vesicle, where the ductus deferens passes the ureter superiorly to reach the medial side of the seminal vesicle (Figure 8.2). The ureters course obliquely through the bladder wall for 1.5 cm.

In the female, the **ureter** (p. 327) can also be followed retroperito-neally along the lateral pelvic wall after it crosses the pelvic brim at the bifurcation of the common iliac artery. After removing the peri-toneum from the lateral pelvic wall, note that the ureter passes in-ferior and posterior to the ovary on the lateral pelvic wall, forming the posterior boundary of the ovarian fossa. Follow the ureter as it leaves the pelvic wall to pass through the endopelvic fascia at the base of the broad ligament to reach the bladder. In this course, it passes 1 to 2 cm lateral to the cervix and uterus. Note that it is crossed superiorly by the uterine artery (Figure 8.3), an important relationship to be considered during pelvic surgery.

The muscular **urinary bladder** lies against the symphysis pubis anteriorly and rests on the pelvic diaphragm inferiorly. As was de-scribed earlier, the bladder is covered superiorly by the peritoneum and when it is distended, the superior surface elevates into the ab-dominal cavity.

Because of its relationships to the pubic bones, the bladder is de-scribed as having a base (fundus), neck, apex, superior surface and two inferolateral surfaces. The **base,** or fundus, faces posteriorly and is closely related to the anterior vaginal wall in the female and the seminal vesicle and rectum in the male. The **neck** is the lowest aspect of the bladder and is pierced by the urethral orifice. The **apex** is the anterior, superior part of the bladder and is attached to the

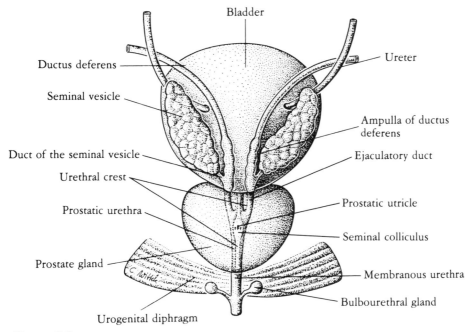

Bladder

Ductus deferens

Seminal vesicle

Ureter

Duct of the seminal vesicle

Urethral crest

Ampulla of ductus deferens

Ejaculatory duct

Prostatic urethra

Prostatic utricle

Seminal colliculus

Prostate gland

Membranous urethra

Bulbourethral gland

Urogenital diphragm

Figure 8.2

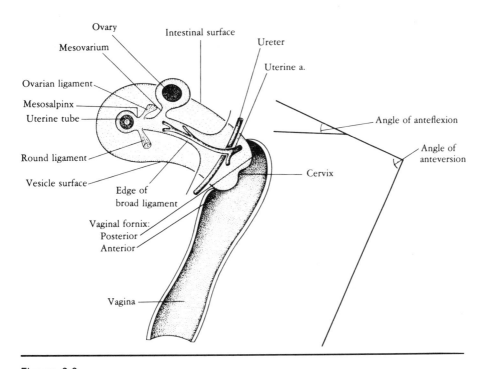

Ovary

Intestinal surface

Mesovarium

Ureter

Uterine a.

Ovarian ligament

Mesosalpinx

Uterine tube

Angle of anteflexion

Angle of anteversion

Round ligament

Vesicle surface

Cervix

Edge of broad ligament

Vaginal fornix:
Posterior
Anterior

Vagina

Figure 8.3

urachus. The **superior surface** extends from the apex anteriorly to the point where the ureters enter the bladder posteriorly. In both the male and the female, this surface is covered by the peritoneum and coils of the intestines. In the female, it is also covered by the uterus and vesicouterine pouch. The two **inferolateral surfaces** are related to the pubic bones.

Internally, observe that the mucosa of the bladder has many irregular folds, except at the base, where there is a smooth, triangular, posterior area called the **trigone.** The **ureteral orifices** are at the superior angles of the trigone and the **internal urethral orifice** is at the apex of the trigone. Review the autonomic innervation of the bladder and its function in relation to the detrusor muscle during micturition. Its blood supply will be discussed later.

72. PELVIC VISCERA
A. Male

If a detailed study of the testis and contents of the spermatic cord (**ductus deferens, testicular artery, deferential artery** and **pampiniform venous plexus**) has not been made, complete it at this time (pp. 310; 312; 315). Observe the position of the **testis** in the scrotal sac, with the dense **tunica albuginea** covering its surface. Note that the **epididymis** is attached to the posterior surface of the testis adjacent to the point where the blood vessels and nerves enter the testis from the spermatic cord. From the tail of the epididymis, the **ductus deferens** continues into the spermatic cord. Follow its course through the superficial and deep rings of the inguinal canal. At the deep ring, the ductus deferens passes lateral to the inferior epigastric artery. It then passes retroperitoneally at the lateral wall of the pelvis to reach the posterior surface of the bladder. Note that on the bladder, it passes superior to the **ureter** and medial to the **seminal vesicle** (p. 316). At the base of the prostate gland, it joins the **duct of the seminal vesicle** to form the **ejaculatory duct** (Figure 8.2). Identify the two finger-shaped seminal vesicles lying in the fascia on the fundus and posterior surface of the bladder.

Observe the **prostate gland** lying inferior to the bladder, resting on the urogenital diaphragm inferiorly. The **base** of the prostate is superior and its **apex** is inferior at the urogenital diaphragm (Figure 8.2). The abundant, dense layer of connective tissue on the gland fuses with the bladder. Also, some of the muscle fibers of the urogenital diaphragm blend with it. Posteriorly, the gland is covered by the peritoneum and is adjacent to the rectum, with the rectovesical pouch and septum between them. The prostate gland encloses the first part (prostatic) of the male urethra and the ejaculatory ducts (Figure 8.2). Observe the courses of the urethra and ducts on the median section. Internally, **right** and **left lateral lobes** of the prostate are anterior to the urethra and a **median lobe** is found superiorly between the urethra and the ejaculatory ducts. A **posterior lobe** is found inferior to the ejaculatory ducts. These lobes can be detected best on the median surface. Note the course of the prostatic urethra through the gland (Figure 8.2) and study the structures on the posterior wall of the prostatic urethra where the ejaculatory ducts enter (p. 312). Identify the **seminal colliculus, urethral crest, prostatic utricle** and **ejaculatory duct.** In the male, the urethra is divided into **prostatic, membranous** and **spongy** (penile) **parts** based on its course through the prostate gland, urogenital diaphragm and corpus spongiosum, respectively (p. 311). Review the characteristics of each of these structures in your textbook and on the median section.

Observe the two **seminal vesicles** on the posterior surface of the bladder. These sacs are lateral to the ampullae of the two ductus deferens and taper inferiorly, where they join the ductus deferens to form the ejaculatory duct. They are separated from the rectum by the rectovesical pouch. Remove the fascia and peritoneum from their surfaces and review their relationships to the bladder, ureters and prostate gland.

B. Female

Observe the **ovary** (p. 327), which is located on the lateral pelvic wall in the **ovarian fossa.** The fossa is bounded inferiorly by the **ureter** and superiorly by the **external iliac vessels.** The long axis of the ovary is vertical. Note the fimbriated end of the uterine tube that covers the superior pole. Observe also that the **suspensory ligament of the ovary** is attached to the upper pole. Identify the **ovarian vessels** within the suspensory ligament. The **ovarian ligament** connects the inferior pole of the ovary to the uterus.

Follow the course of the **uterine tube** between the ovary and the uterus, noting its position in the broad ligament, which covers the surface of the uterine tube. The tube takes a tortuous course and expands to form the **ampulla** before forming the funnel-shaped **infundibulum** at the ovary.

The **uterus** (pp. 329–331) is the thick-walled, muscular organ in which the fertilized ovum implants and the embryo develops. It is shaped like a pear and has a narrow internal cavity. Most of the uterus is horizontal, as will be described. Identify its three parts (p. 331):

1. **Fundus.** This is the rounded distal end superior to the uterine tubes.

2. **Body.** This largest part of the uterus receives the uterine tubes and narrows inferiorly as an **isthmus,** which continues as the **cervix.**

3. **Cervix.** The neck of the uterus that projects into the vagina is the cervix. The cavity (canal) of the cervix opens into the vagina at the **external os.** Observe these structures on the median section.

Because the uterus (p. 329) is positioned horizontally, the superior surface of the body is the **intestinal surface** and the inferior surface is the **vesical surface** (Figure 8.3). The cervix is strongly attached to the bladder and is the least mobile. However, superior to the cervix, the body and fundus are mobile. The normal anterior angulation of the uterus between the body and the cervix is the **anteverted position;** the angulation between the body and the fundus is the **anteflexed** position (Figure 8.3). Note that these positions are movable. The uterus is supported by the **pelvic diaphragm, urogenital diaphragm** and **perineal membrane.** The adjacent bladder and rectum also offer support. In addition, the **perineal body** provides strong support for the uterus. The blood supply to the uterus will be dissected with the iliac vessels.

Observe that the **vagina** (pp. 329–332) forms an angle of 90 to 110 degrees with the long axis of the uterus and cervix (Figure 8.3). The vagina is the terminal part of the reproductive tract and opens externally at the vaginal orifice into the vestibule. Note that the cervix projects into the anterior wall of the vagina, making it shorter than the posterior wall. Around the vaginal surfaces of the cervix, a cul-de-sac is formed and is divided into the **anterior, posterior** and **lateral vaginal fornices.** Anteriorly, the vagina is attached to the fundus of the bladder above and is tightly fused to the urethra below. Posteriorly, it is related to the rectouterine pouch, rectum and perineal body.

The **female urethra** (pp. 326; 329) is much shorter than the male urethra. After leaving the bladder, the urethra pierces the sphincter urethrae muscle of the urogenital diaphragm. It then courses anterior to the lower half of the vagina and is fused tightly to it. Externally, the urethra opens into the vestibule between the vagina and the clitoris.

73. PELVIC FLOOR AND WALLS
A. Muscles and Fasciae

Review the bony framework that forms the boundaries of the lesser (true) pelvis. Remove completely the peritoneum lining the lateral walls of the pelvis to procede with the following. The lateral and posterior walls of the true pelvis are covered by the **obturator internus** and **piriformis** muscles, respectively. The pelvic floor is enclosed by the muscular sling referred to as the **pelvic diaphragm** (Figure 8.4).

1. Piriformis. Observe the **piriformis** muscle (p. 319), which arises from the pelvic surface of the sacrum around the second, third and fourth sacral foramina (Figure 8.4). It is triangular and its apex passes through the greater sciatic foramen to enter the gluteal region, where it inserts on the greater trochanter of the femur. This insertion and the latter course of the piriformis are observed in the dissection of the gluteal region. Mobilize the rectum and observe the piriformis deeply on the sacrum. Note that the sacral plexus of nerves is located on the pelvic surface of the piriformis.

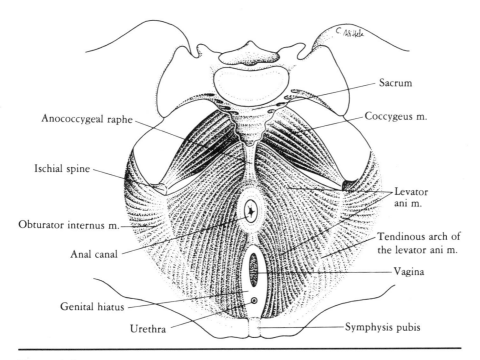

Figure 8.4

2. Obturator Internus. The **obturator internus** muscle (p. 319) covers the lateral wall of the minor pelvis (Figure 8.4). Identify its fibers, which surround the margins of the **obturator foramen.** Many of the fibers of this muscle arise from the obturator membrane, which encloses the obturator foramen, except for the **obturator canal** superiorly. Follow the muscle fibers as they pass posteriorly out of the pelvis through the lesser sciatic foramen. At this point, the muscle makes a 90-degree turn and enters the gluteal area to insert on the greater trochanter, where it is dissected. Note that a dense layer of **obturator fascia** covers the muscle on its medial surface. A thickening of this fascia forms the **tendinous arch of the levator ani** (Figure 8.4), which crosses between the superior pubic ramus and the ischial spine. As will be described, some of the fibers of the levator ani arise from this tendinous arch. Observe the courses of the **obturator nerves** and **vessels** as they cross this muscle medially to reach the obturator foramen and canal.

3. Pelvic Diaphragm. The pelvic diaphragm encloses the inferior outlet of the pelvis and is often referred to as forming the pelvic floor. The diaphragm is covered on its inferior surface by the inferior fascia of the pelvic diaphragm, which is continuous with the obturator internus fascia in the ischiorectal fossa. On its superior surface, the diaphragm is covered by the superior fascia of the pelvic diaphragm, which is continuous with the transversalis fascia. The diaphragm is shaped like a funnel and extends from the symphysis pubis anteriorly to the coccyx posteriorly. In the midline, the two halves of the diaphragm are separated by the **genital hiatus,** through which parts of the urinary, reproductive and digestive systems pass. The muscle fibers of the pelvic diaphragm may be extremely thin, often consisting mostly of a membranous layer of connective tissue that contains a few muscle fibers. The pelvic diaphragm is divided into two parts, the **levator ani** and **coccygeus** muscles (Figure 8.4).

The **levator ani** (pp. 316; 319) is the anterior part of the diaphragm and is divided into three portions: the puborectalis, pubococcygeus and iliococcygeus muscles. It may be difficult to differentiate these muscles and it is usually best to examine the levator ani as a single muscle. In the minor cavity, review and identify the **dorsal (pelvic) surface of the symphysis pubis** and the **ischial spine.** Mobilize the viscera and note that the anterior fibers of the levator ani arise from the dorsal pubis. These fibers (puborectalis) border the genital hiatus and form a U-shaped sling around the rectum. The fibers of the iliococcygeus arise from the **tendinous arch of the levator ani** (as previously noted) on the pelvic surface of the obturator internus and insert on the coccyx. The point of origin of the iliococcygeus separates the upper half of the obturator internus in the pelvis from the lower half of this muscle in the ischiorectal fossa.

4. The **coccygeus** (p. 319) is a thin, triangular muscle. Note that its apex attaches to the ischial spine with its base attaching to the sacrum and coccyx. It completes the muscular floor of the pelvis between the levator ani and the piriformis. The **sacrospinous ligament** is immediately external to this muscle.

B. Nerves

The nerves of the pelvis are the **sacral plexus** and the inferior components of the **autonomic nervous system** (Figure 8.5). The sacral plexus is primarily motor and cutaneous to the walls of the lower abdomen, perineum, pelvis and most of the lower limb. Note that it is formed on the surface of the piriformis (p. 319). Review the lumbar plexus and identify the **lumbosacral trunk** (ventral rami of the fourth and fifth lumbar spinal nerves). Inferior to the trunk, identify the ventral rami of the **first, second, third** and **fourth sacral spinal nerves.** The fifth nerve is extremely small. The first four

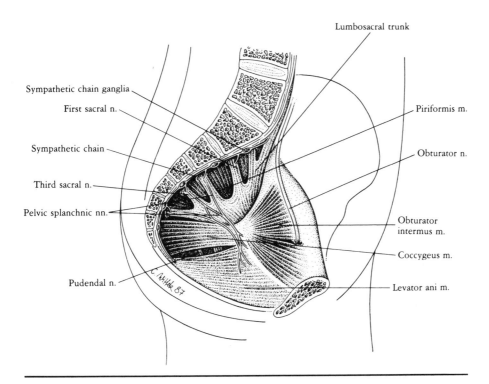

Figure 8.5

sacral nerves and the lumbosacral trunk form the sacral plexus. Most of its branches leave the pelvis through the greater sciatic notch. Many of these terminal branches are dissected during the study of the gluteal compartment of the lower limb and perineum. Review and identify the **superior** and **inferior gluteal nerves** and **pudendal nerve** (p. 319). The superior gluteal nerve leaves the pelvis superior to the piriformis and the inferior gluteal nerve exits the pelvis inferior to this muscle. The pudendal nerve leaves the pelvis by passing deep to the superior margin of the coccygeus. Study the complete description of the plexus in your textbook.

The autonomic fibers (p. 308) of the pelvis are contained in the following structures, each of which should be identified:

1. **Sympathetic trunks.** The terminal ends of the two sympathetic trunks (Figure 8.5) enter the pelvis on the sacrum and contain a few small sympathetic ganglia.

2. **Superior** and **inferior hypogastric plexuses.** The superior hypogastric plexus is at the sacral promontory at the bifurcation of the aorta; the inferior hypogastric plexus (pelvic plexus) is on each side of the rectum in the pelvis. Most of the fibers that enter the inferior hypogastric plexus are either sympathetic fibers, which are derived mainly from the lower **lumbar splanchnic nerves** that pass via the superior plexus, or parasympathetic fibers, which are derived from the **pelvic splanchnic nerves** (Figure 8.5).

3. **Pelvic splanchnic nerves.** The fibers of these nerves enter the inferior hypogastric plexus. They are preganglionic, parasympathetic fibers that are derived from the second, third and fourth sacral spinal cord segments. Identify these strands of nerves (Figure 8.5) as they pass from the sacral ventral rami to the inferior hypogastric plexuses on either side of the rectum. The fibers of the pelvic splanchnic nerves provide parasympathetic innervation to the genitalia, all the pelvic viscera and descending and sigmoid colon.

From the pelvic plexus, sympathetic and parasympathetic fibers are provided to the various pelvic viscera. Review these pathways in your textbook. Visceral afferent fibers distribute with fibers of the inferior plexus.

C. Vessels

The blood supply (Figure 8.6) of the pelvis and most of the lower limb is provided by the distributions of the two **common iliac arteries** (pp. 318–319). Note that these two arteries arise from the bifurcation of the abdominal aorta at the level of the fourth lumbar vertebra on the posterior abdominal wall. Follow them along a straight line for a

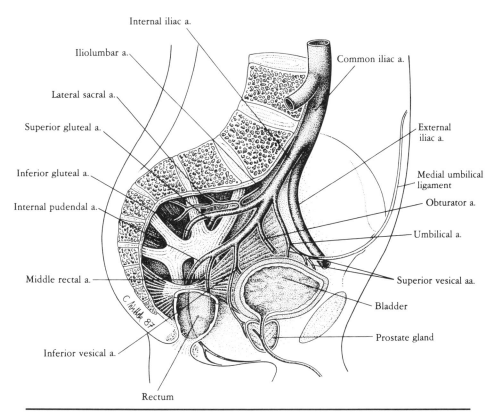

Internal iliac a.

Iliolumbar a.

Lateral sacral a.

Superior gluteal a.

Inferior gluteal a.

Internal pudendal a.

Middle rectal a.

Inferior vesical a.

Rectum

Common iliac a.

External iliac a.

Medial umbilical ligament

Obturator a.

Umbilical a.

Superior vesical aa.

Bladder

Prostate gland

Figure 8.6

short distance at the pelvic brim. The common iliac arteries divide into the **internal** and **external iliac arteries** (Figure 8.6). Recall that the ureter crosses the pelvic brim at the bifurcation of the common iliac arteries. The **external iliac artery** is the direct continuation of the common iliac artery. Follow it along the **psoas major** as it courses deep to the inguinal ligament, where it enters the thigh to become the **femoral artery.** Its last branch at the inguinal ligament is the **inferior epigastric artery.** Review the relationships of this artery to the deep inguinal ring and direct and indirect inguinal hernias. Locate the **external iliac vein** medial to the artery and review its course.

The **internal iliac artery** (pp. 318–319) descends into the pelvis along its lateral wall. It supplies blood to the pelvic viscera, perineum, gluteal region and medial parts of the thigh. Note that it is a short, thick vessel. It usually divides into anterior and posterior branches, although there are many variations in the branching pattern of the artery. Identify the following parietal and visceral arteries (Figure 8.6) (pp. 318–319), noting that many of the branches course with similarly named nerve branches of the sacral plexus:

1. **Iliolumbar artery.** This artery ascends superiorly to the iliac fossa to supply muscles of the posterior abdominal and pelvic walls (iliacus, psoas major and quadratus lumborum).

2. **Lateral sacral artery.** This artery passes medially on the pelvic surface of the sacrum and provides branches to the sacral canal and muscles on the sacrum.

3. **Superior gluteal artery.** This artery usually passes with the superior gluteal nerve between the lumbosacral trunk and the first sacral ventral ramus or between the first and second sacral rami. It leaves the pelvis through the greater sciatic foramen and is dissected further in the gluteal region.

4. **Obturator artery.** This artery crosses the lateral pelvic wall with the obturator nerve to enter the thigh through the obturator canal. Observe that it is crossed by the ureter and ductus deferens in the male and by the ureter in the female.

5. **Internal pudendal artery.** This artery is one of the terminal branches of the anterior division and often arises in common with the inferior gluteal artery. It leaves the pelvis by passing just superior to the coccygeus with the pudendal nerve. The later courses of the artery and nerve are dissected in the gluteal and perineal regions.

6. **Inferior gluteal artery.** This artery is larger than the internal pudendal artery and passes between the rami of the sacral plexus to exit the pelvis inferior to the piriformis with the inferior gluteal nerve to enter the gluteal area. The inferior gluteal artery can branch in common with the superior gluteal or internal pudendal artery.

7. **Umbilical artery.** This artery is the first visceral branch. Note that it is short and passes toward the bladder, where it gives rise to the **superior vesical arteries** to the bladder and a small artery to the ductus deferens. Distal to this point, the artery atrophies and becomes the **medial umbilical ligament.**

8. **Inferior vesical** and **middle rectal arteries.** These two visceral arteries can form separately or as a common branch. The **inferior vesical artery** supplies the deep surface of the bladder in both sexes as well as the prostate and seminal vesicle in the male. In the female, it often forms vaginal branches. The **middle rectal artery** passes to the middle third of the rectum, where it anastomoses with the superior and inferior rectal arteries. Follow these arteries to their respective viscera.

9. **Uterine artery.** The uterine artery (p. 332) usually forms as a separate branch of the anterior division. Identify this artery as it

crosses the pelvic floor in the base of the broad ligament. It reaches the uterus at the level of the cervix. Lateral to the uterus, the artery passes superior to the ureter and then forms tortuous ascending and descending branches along the body of the uterus and vagina (Figure 8.3). Additional **vaginal branches** may arise from the inferior vesical artery or directly from the anterior division of the internal iliac artery.

Identify the **internal iliac vein** medial to the artery. It has tributaries that correspond in name to the branches of the internal iliac artery.

CHAPTER NINE

HEAD AND NECK

74. SURFACE ANATOMY OF THE NECK

The dissection of the posterior aspect of the neck should now be reviewed (see Chapter 2). To examine the surface anatomy of the anterior neck, palpate the **mental protuberance** of the mandible anteriorly at the chin. Then, with your fingers, descend from the chin down the midline of the neck to palpate the **hyoid bone** at the junction of the soft structures that form the floor of the mouth (suprahyoid area) with the neck. Grasp the hyoid bone and move it from side to side. It lies at the transverse level of the third cervical vertebra. Immediately inferior to the hyoid bone, palpate the **thyroid cartilage** with its **prominence** (Adam's apple) and **lateral laminae.** The thyroid cartilage is at the transverse level of the fourth and fifth cervical vertebrae. With your finger, carefully feel the narrow space inferior to the thyroid cartilage and locate the **cricoid cartilage.** This ring-shaped cartilage is firm and at the level of the sixth cervical vertebra. Inferior to the cricoid cartilage, palpate the cartilaginous rings of the trachea inferiorly to a level 2 to 3 cm superior to the suprasternal notch (p. 144).

Laterally, identify the **sternocleidomastoid** muscle, which courses diagonally from the manubrium sterni and clavicle anteriorly and inferiorly to the mastoid process posteriorly and superiorly. This muscle forms a lateral bulge in the neck and is most prominent as one turns the head. Posterior to the muscle, the lateral aspect of the neck is limited by the clavicle inferiorly. Major cutaneous branches of the cervical plexus (lesser occipital, great auricular, transverse cervical and supraclavicular nerves) enter the skin at the junction of the middle and upper thirds of the posterior edge of the sternocleidomastoid to supply the skin of the lateral neck, lower face and upper chest. The external jugular vein (pp. 158; 161), is often visible on the external surface of the sternocleidomastoid and passes inferiorly deep to the clavicle. Deep to the inferior fibers of the muscle, the subclavian artery and vein and trunks of the brachial plexus also course deep to the clavicle. Anterior to the muscle at the superior margin of the thyroid cartilage, the common carotid artery divides into internal and external carotid arteries.

75. SKIN REFLECTIONS AND SUPERFICIAL STRUCTURES OF THE NECK

With the cadaver in the supine position, make the skin incisions shown in Figure 9.1. Be careful to preserve the subcutaneous structures (cutaneous vessels and nerves and platysma muscle described below) as the skin is reflected (Figure 9.2). Make a vertical, midline

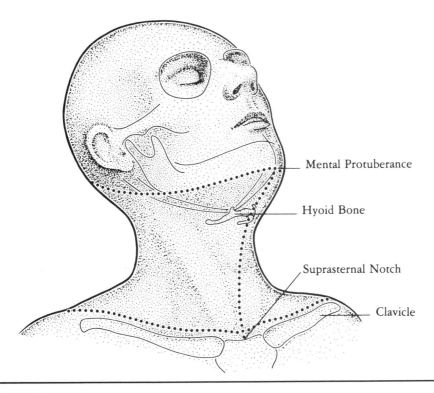

Figure 9.1

incision from the mental area of the mandible inferiorly to the su-prasternal notch. From the superior end of this incision, make a transverse cut posteriorly along the inferior border of the mandible to the mastoid area posterior to the earlobe. Make another transverse cut posteriorly along the clavicle from the inferior end of the midline incision at the suprasternal notch. Reflect this layer of skin laterally to meet the skin incisions previously made on the posterior neck. Be careful to identify and preserve the subcutaneous **platysma** muscle (pp. 58; 165) overlying the clavicle, which should now be carefully loosened inferiorly and reflected superiorly to the mandible. The pla-tysma is a muscle of facial expression and is a broad sheet that covers the lateral surface of the neck. Reflecting this muscle superiorly to the mandible is one of the most difficult parts of the dissection of the superficial neck. Clean and identify the anterior and posterior bor-ders of the platysma. Beginning at its inferior border, blunt dissect the platysma from the deeper fascia. As the platysma is pulled superiorly, protect the cutaneous nerves (**transverse cervical, supraclavicular** and **great auricular nerves**), **external jugular vein** and **cervical branch of the facial nerve,** which innervates

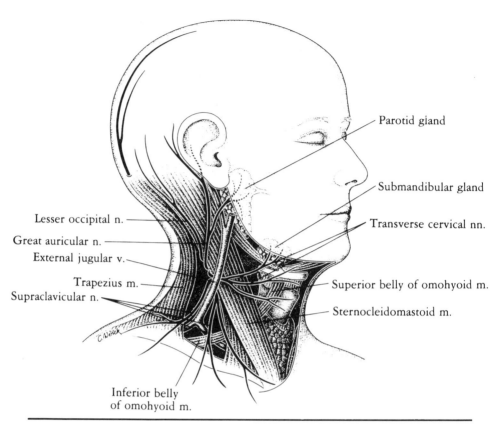

Parotid gland

Submandibular gland

Transverse cervical nn.

Lesser occipital n.

Great auricular n.

External jugular v.

Trapezius m.

Supraclavicular n.

Superior belly of omohyoid m.

Sternocleidomastoid m.

Inferior belly
of omohyoid m.

Figure 9.2

the platysma (Figure 9.2) (p. 161). Although the platysma is difficult to separate from the deeper fascia, it is essential that you completely reflect it superiorly to the inferior border of the mandible.

Review the layers of the cervical fascia and carefully remove the superficial layer that covers the neck. Note many of the previously described cutaneous nerves and vessels that course superficially in the tela subcutanea. Identify and clean completely the anterior and posterior borders and lateral surface of the **sternocleidomastoid** superiorly to the mastoid process (p. 161) and preserve the following structures. Trace the **transverse cervical, supraclavicular, lesser occipital** and **great auricular cutaneous nerves** from the posterior border of the muscle through the tela subcutanea of the superficial neck (Figure 9.2). These nerves are cutaneous branches of the cervical plexus. Study the distributions of these nerves in your textbook. The origins of these branches from the cervical plexus will be dissected later. The lesser occipital nerve is inconspicuous and can be dissected better at a later time.

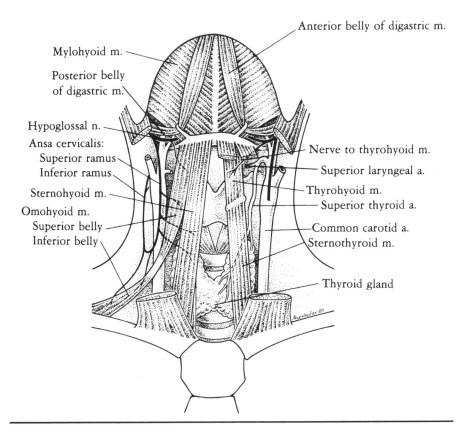

Mylohyoid m.

Posterior belly
of digastric m.

Hypoglossal n.
Ansa cervicalis:
  Superior ramus
  Inferior ramus
Sternohyoid m.
Omohyoid m.
  Superior belly
  Inferior belly

Anterior belly of digastric m.

Nerve to thyrohyoid m.
Superior laryngeal a.
Thyrohyoid m.
Superior thyroid a.
Common carotid a.
Sternothyroid m.

Thyroid gland

Figure 9.3

In the anterior midline, dissect the **anterior jugular veins.** These
veins and the **external jugular veins** (p. 160) are deep to the platysma at the lateral and anterior sides of the neck. The anterior
jugular vein communicates superiorly with the submental veins at
the floor of the mouth. Inferiorly, the anterior jugular vein drains
into the external jugular vein through transverse communicating
veins that course deep to the sternocleidomastoid muscle 3 cm superior to the manubrium sterni and clavicle. The external jugular
vein is formed by the confluence of the retromandibular and posterior
auricular veins posterior to the angle of the mandible. This origin of
the jugular vein will be studied with the parotid gland. Locate the
external jugular vein as it descends superficial to the sternocleidomastoid. During its course, it receives the posterior external jugular, suprascapular and transverse cervical veins. The external jugular vein terminates into the subclavian vein in the posterior cervical
triangle (p. 161) by piercing the superficial layer of the cervical fascia
superior to the clavicle.

76. ANTERIOR TRIANGLE

The anterior triangle is bounded by the inferior border of the mandible, anterior border of the sternocleidomastoid and anterior midline of the neck. This triangle is divided into the muscular, carotid and submandibular triangles. The small submental triangle, which is of lesser importance, is also described. Identify the positions and courses of the **anterior** and **posterior bellies of the digastric** muscle. Clean them from their fascia. These two bellies along with the inferior border of the mandible form the boundaries of the **submandibular (suprahyoid)** division of the anterior triangle. Inferior to the hyoid bone, observe the omohyoid muscle, which passes obliquely across the midpoint of the anterior triangle. The omohyoid divides the anterior triangle into the **carotid triangle,** which is bordered by the omohyoid, posterior belly of the digastric and anterior border of the sternocleidomastoid muscles, and the **muscular triangle,** which is bordered by the omohyoid, anterior border of the sternocleidomastoid and anterior midline of the neck. Review the boundaries of these triangles in your textbook and atlas (p. 145). You will now dissect these triangles except the submandibular triangle, which will be studied later in association with the face and floor of the mouth.

A. Muscular Triangle

Review the midline positions of the **hyoid bone, thyroid** and **cricoid cartilages** and **trachea** (Figure 9.3), noting their relationships to the muscular triangle (p. 144). Other contents of this triangle include the infrahyoid (strap) muscles and the thyroid and parathyroid glands. The four **infrahyoid** muscles extend from the clavicle, scapula and manubrium inferiorly to the hyoid bone and thyroid cartilage superiorly (pp. 144; 160–165). They form two muscular planes that lie ventral to the larynx, trachea and thyroid gland (Figure 9.3). These muscles are important in movements of the hyoid bone and thus in movements of the tongue, swallowing and phonation. As the muscles are cleaned of their fascia, watch for the small nerves that approach laterally to reach the deep surfaces of these muscles. These small nerves will be used later to identify their origin from the ansa cervicalis.

The two superficial strap muscles are the **sternohyoid** and **omohyoid** (p. 161). The sternohyoid is more medial, narrow and almost vertical in orientation. The omohyoid is more lateral and consists of two bellies. Identify the sternohyoid and omohyoid muscles and remove the muscular layer of the cervical fascia covering them. The anterior jugular vein should be reflected ventral to them. The **su-**

perior belly of the omohyoid attaches to the hyoid bone and descends deep to the sternocleidomastoid, where it is attached to the clavicle by a fascial sling. From this point, the inferior belly descends across the posterior triangle to attach to the superior border of the scapula medial to the scapular notch. These two superficial muscles are innervated by branches of the ansa cervicalis.

The deeper layer of strap muscles is formed by the **sternothyroid and thyrohyoid** muscles (pp. 163; 165). The more superficial sternohyoid muscle described previously should now be cut and reflected at its midpoint to expose the deeper layer. These deeper two muscles form a single plane of fibers. Identify the **thyrohyoid** deep to the superior fibers of the superficial group and the wide **sternothyroid** deep to the inferior part of the sternohyoid. The thyroid gland lies deep to the sternothyroid. The sternothyroid is innervated by the ansa cervicalis and the thyrohyoid is supplied by the nerve to the thyrohyoid. The latter nerve consists of fibers that are derived mainly from the first and second cervical nerves that are carried via the hypoglossal cranial nerve and will be dissected in the carotid triangle.

Identify and clean the borders of the **thyroid gland** (pp. 160; 164; 172) after reflecting the sternothyroid at its midpoint. Note that the **lateral lobes** of the gland ascend to the oblique line of the thyroid cartilage deep to the insertion of the sternothyroid. Inferiorly, the lateral lobes may descend to the sixth tracheal ring. The gland embraces the trachea medially and its lateral lobes are connected by an **isthmus** that crosses the midline at the second, third and fourth tracheal rings. A **pyramidal lobe** can often be found extending superiorly, usually from the left lobe. The gland is related anteriorly to the strap muscles, laterally to the carotid sheath and medially to the trachea and recurrent laryngeal nerve (p. 173). Identify these relationships. Cut the isthmus and reflect the gland to expose the recurrent laryngeal nerve deep to the thyroid gland as it courses between the trachea and esophagus. The gland is supplied by the **superior thyroid artery,** which can be followed deep to the strap muscles to reach the superior lobe (Figure 9.3). The superior thyroid artery arises from the external carotid artery. The inferior thyroid artery supplies the gland and is deep to the inferior edge of the lateral lobe and will be dissected later in the root of the neck. The superior, middle and inferior thyroid veins drain the thyroid gland. Review this pattern of venous drainage. Dissect these vessels in more detail later when the carotid sheath is exposed.

Although the parathyroid glands are difficult to find in the laboratory, two to six small masses are located deep to the thyroid fascia on the posterior aspect of the thyroid gland.

B. Carotid Triangle

Dissect and expose the superior boundary of this triangle, which is formed by the **posterior belly of the digastric** muscle (p. 165), by elevating the large, superficial part of the **submandibular gland** that overlies the carotid triangle covering the posterior digastric belly (p. 163). Identify the stylohyoid muscle and note its relationship to the posterior digastric muscle. The posterior belly of the digastric muscle forms an important landmark and should be carefully exposed. Do not dissect superior to the posterior belly into the submandibular region because this area will be studied later.

Dissection of the carotid triangle (Figure 9.4) is important because this region contains the major parts of the external carotid artery system, the internal jugular vein and parts of the courses of the vagus and hypoglossal cranial nerves. Clean and expose completely

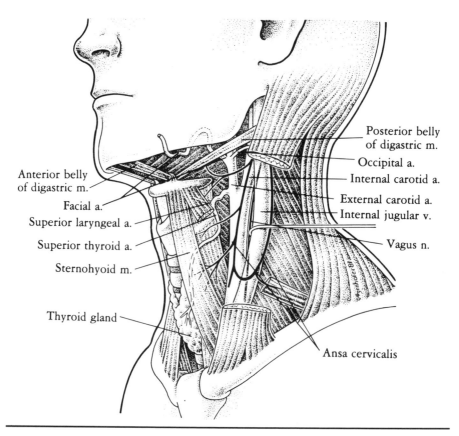

Figure 9.4

the anterior and posterior borders of the sternocleidomastoid if this dissection has not been done. To gain access to the structures of the carotid triangle, bisect the sternocleidomastoid (p. 163) at its midpoint. Retract the cut ends superiorly to the mastoid process and inferiorly to the sternum. Carefully tease the muscle from the carotid sheath and other deep structures. The **carotid sheath** is a fascial covering of the common and internal carotid arteries, internal jugular vein and vagus nerve. The sheath extends from its attachment to the base of the skull inferiorly to the superior thoracic aperture.

Before stripping away the carotid sheath, it is critical to identify the **ansa cervicalis** (Figure 9.4), which is embedded in the sheath (pp. 163; 166; 168–169). The ansa is a loop of nerve fibers that consist of superior and inferior roots derived from branches of the cervical plexus (C1 to C3). The ansa cervicalis encircles and is embedded in the carotid sheath. The fibers of the ansa innervate the omohyoid, sternohyoid and sternothyroid. By following the small nerves identified earlier on the deep surface of the strap muscles back toward the carotid sheath, the **superior** and **inferior roots** of the ansa cervicalis can be separated from the carotid sheath. The superior root fibers (C1 and C2) travel with the hypoglossal nerve for a short distance before dropping off to form the superior root of the ansa cervicalis in the carotid triangle. After identifying the superior root on the sheath, bluntly dissect it superiorly toward the posterior digastric muscle and observe it branching from the **hypoglossal nerve** (pp. 168–169). The inferior root of the ansa cervicalis contains fibers of the second and third cervical ventral rami. These fibers usually are lateral to the carotid sheath where they join the superior root. The inferior root will be followed later to its origin from the cervical plexus during the dissection of the posterior triangle. Some of the fibers of C1 and C2 continue distally with the hypoglossal nerve to supply the thyrohyoid and geniohyoid but usually are not considered to be part of the ansa cervicalis.

Open the carotid sheath in the carotid triangle and identify the **common carotid artery, internal jugular vein** and **vagus nerve** (Figure 9.4) (pp. 167; 170). The right common carotid artery branches from the brachiocephalic trunk deep to the sternoclavicular joint. The left common carotid artery branches directly from the aortic arch within the superior mediastinum and thus is longer than the right common carotid artery. The inferior course of the common carotid arteries in the neck will be dissected in the posterior triangle. Follow the common carotid arteries superiorly to the superior border of the thyroid cartilage, where they divide into the **internal** and **external carotid arteries.** The internal carotid artery courses in the carotid sheath superiorly to the base of the skull, where it enters the cranial cavity through the carotid canal to supply the brain and orbit. There are no branches from the internal carotid artery in the neck.

Within the carotid triangle, the **external carotid artery** gives rise to four or five of its eight branches (Figure 9.4). It then ascends deep to the posterior belly of the digastric muscle to enter the retromandibular region. From the anteromedial aspect of the external carotid artery in the carotid triangle, locate the **superior thyroid, lingual** and **facial arteries** (pp. 156–157; 165; 167; 170). Clean the fascia from these arteries and review their relationships. Note that the superior thyroid artery is usually the first branch at the level of the carotid bifurcation. It descends deep to the strap muscles into the muscular triangle to supply the thyroid gland and associated muscles. From the superior thyroid artery, identify the **superior laryngeal artery** (p. 165), which passes with the **internal branch of the superior laryngeal nerve** through the **thyrohyoid membrane** to enter the larynx superior to the thyroid cartilage. Note the relationships of this vessel and nerve to the posterior border of the **thyrohyoid** muscle. Watch for the small **external branch** of the superior **laryngeal nerve** as it courses with the superior thyroid artery deep to the strap muscles to innervate the cricothyroid muscle. The **lingual** and **facial arteries** arise at the level of the hyoid bone (often as a common artery). They have a short course in the carotid triangle and then ascend deep to the posterior belly of the digastric muscle to enter the submandibular (suprahyoid) region, where their distributions and relationships will be studied later.

The ascending pharyngeal artery branches from the posterior aspect of the external carotid artery within the carotid triangle and enters the pharynx after ascending along its lateral aspect. The ascending pharyngeal artery is small and on the deep aspect of the external carotid artery, making it difficult to locate at this stage of the dissection. It is better identified during a later posterior approach to the pharynx. The **occipital artery** is a branch of the posterior aspect of the external carotid artery at the level of the lingual origin, and it passes laterally along and deep to the inferior border of the posterior digastric muscle. The origin of the occipital artery may or may not be considered within the carotid triangle. Observe the **hypoglossal nerve** in the carotid triangle and note that it characteristically loops inferior to the sternocleidomastoid branch of the occipital artery. The hypoglossal nerve crosses the carotid sheath and lingual artery laterally to enter the submandibular triangle.

Study the courses of the internal jugular vein and its tributaries.

In the carotid triangle, identify the **vagus nerve** (pp. 167; 170) as it descends in the carotid sheath between the internal jugular vein and the carotid arterial system (Figure 9.4). There are several important branches from the vagus nerve in the neck to the pharynx and larynx. The pharyngeal branches of the vagus are not in the carotid triangle and will be dissected later with the posterior aspect of the pharynx. The superior laryngeal nerve also arises from the vagus nerve on the posterior pharynx before the vagus enters the carotid triangle. Its origin will also be seen better with the posterior pharynx. However, two branches of the superior laryngeal nerve are dissected in the carotid triangle. The superior laryngeal nerve descends and divides posterior to the carotid sheath into two branches that enter the carotid triangle: the **internal** and **external branches of the superior laryngeal nerve.** The external branch is small and follows the course of the superior thyroid artery to the muscular triangle to innervate the **cricothyroid** muscle. The internal branch was previously identified traveling with the superior laryngeal artery as it enters the larynx at the thyrohyoid muscle. Several **cardiac branches** of the vagus nerve can also be recognized passing medially and inferiorly to reach the superior thoracic aperture.

Elevate the carotid artery to expose the **sympathetic trunk** (pp. 170; 172), which lies on the prevertebral fascia and muscles posterior to the carotid sheath. The sympathetic trunk is the superior extension of the thoracic sympathetic trunk through the thoracic aperture. Usually, there are three cervical sympathetic chain ganglia located along this cervical portion of the trunk. These ganglia will be dissected with the posterior pharynx.

77. POSTERIOR TRIANGLE

The posterior triangle (Figure 9.5) is at the lateral side of the neck and is bounded by the trapezius and sternocleidomastoid muscles and middle third of the clavicle (p. 161). Review the position of the trapezius muscle dissected on the posterior surface of the neck. Identify its anterior border, which forms the posterior boundary of the posterior triangle. The external jugular vein crosses the sternocleidomastoid laterally to enter the triangle and the four cutaneous nerves of the cervical plexus curve around the posterior border of the sternocleidomastoid. These nerves and vessel have been previously described and dissected (p. 163). The major contents of the posterior triangle are located inferior to the inferior belly of the omohyoid muscle. Many authors refer to this lower aspect of the triangle and the adjacent area deep to the sternocleidomastoid as the root of the neck. The structures found in these areas will be dissected together.

The first step in dissecting the posterior triangle is removing the dense prevertebral fascia that covers the floor of the triangle. Its

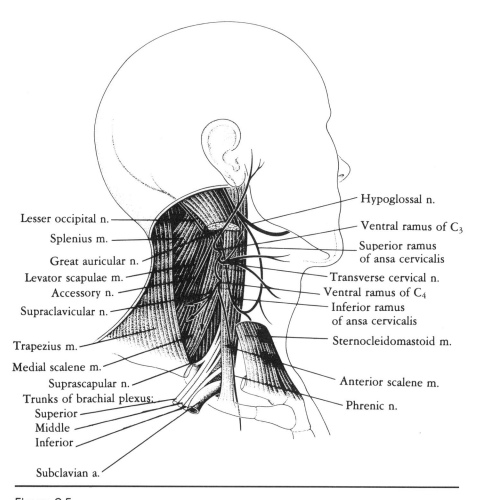

Figure 9.5

removal requires careful attention to avoid damaging the nerves of the cervical plexus located within the floor of the triangle. The most efficient approach in separating this fascia from the nerves is to follow the cutaneous branches of the cervical plexus described earlier toward the floor and isolate the cervical ventral rami and nerves from the fascia as described below.

The **cervical plexus** (pp. 167; 170; 172) is the primary somatic sensory and motor nerve supply to the nonvisceral structures of the neck. The plexus is formed by the first four cervical ventral rami, located on the surface of the scalene muscles. These rami form communicating loops with each other before forming the named nerve components of the cervical plexus that distribute through the posterior triangle. These named nerves include: (1) the four cutaneous nerves previously studied that are formed by the second, third and fourth rami (lesser occipital, great auricular, transverse cervical and supraclavicular); (2) the ansa cervicalis complex (already dissected), which innervates the strap muscles; (3) the phrenic nerve to the diaphragm; (4) the contributions to the accessory nerve; (5) the direct muscular branches; and (6) the gray rami communicantes.

First, review the branches of the supraclavicular cutaneous nerves, which distribute through the posterior triangle and pass over the superficial surfaces of the clavicle and shoulder. The supraclavicular nerves derive from the third and fourth ventral rami of the cervical spinal nerves. Follow the supraclavicular nerves proximally and isolate them through from the fascial floor to identify the **third and fourth ventral rami,** which are embedded in the prevertebral fascia (Figure 9.5). Note that these rami are deep to the reflected sternocleidomastoid muscle and thus are not contents of the posterior triangle. Follow the loops between these two rami. Dissect into the fascia superior to the third ventral ramus to isolate the **second cervical ventral ramus.** Follow the fibers of the second and third rami distally as they converge to form the transverse cervical and great auricular nerves. Also, now follow the **inferior root of the ansa cervicalis** to its origin from the second and third cervical rami.

An additional way to identify the fourth ventral ramus is to isolate the **phrenic nerve** from the prevertebral fascia on the lateral surface of the anterior scalene muscle and follow it superiorly to its main origin from the fourth ventral ramus (Figure 9.5). The phrenic nerve is the motor supply to the diaphragm and is sensory from the pleura and pericardium. The nerve consists mainly of fibers from the fourth cervical ramus with additional contributions from the third and fifth rami. The phrenic nerve descends from lateral to medial across the ventral surface of the anterior scalene muscle deep to the prevertebral fascia (p. 172) to enter the thorax by passing between the subclavian artery and the subclavian vein. This inferior course at the root of the neck will be dissected later. The entire course of the phrenic nerve is deep to the sternocleidomastoid and thus does not enter the posterior triangle.

The direct muscular branches from the cervical plexus are also deep to the sternocleidomastoid and supply the prevertebral muscles of the neck (rectus capitis lateralis, longus capitis and rectus capitis anterior) and scalene muscles.

The floor of the posterior triangle is formed by the **middle** and **posterior scalene, levator scapulae** and **splenius** muscles (pp. 144; 162), which are covered by the prevertebral fascia. The splenius and levator scapulae were dissected earlier and should be reviewed. The scalene muscles, divided into **anterior, middle** and **posterior fibers,** arise from the transverse processes of the cervical vertebrae. The anterior and middle scalene muscles insert into the first rib. Note the vertical gap between these muscles. The posterior muscle, often not clearly separated from the middle scalene muscle, is identified by its insertion on the second rib. The anterior scalene muscle is located deep to the sternocleidomastoid and does not contribute to the floor of the posterior triangle. Clean these muscles from the prevertebral fascia.

The **accessory cranial nerve** (XI) is the most superior, major structure in the posterior triangle (pp. 162–163). It leaves the cranial cavity by way of the jugular foramen and then passes the internal jugular vein laterally to penetrate the sternocleidomastoid muscle. Find this superior course of the accessory nerve by retracting the upper part of the sternocleidomastoid and exposing the nerve embedded on its deep surface lateral to the internal jugular vein. The accessory nerve then enters the posterior triangle, leaving the sternocleidomastoid at the junction of its middle and upper thirds. In the posterior triangle, the nerve crosses the surface of the levator scapulae (Figure 9.5) to pass deep to the trapezius. The accessory nerve innervates the trapezius and sternocleidomastoid muscles, which also receive sensory fibers from the third and fourth cervical nerves.

Inferior to the fourth cervical ventral ramus, the origins of the **brachial plexus** (supraclavicular part) can now be observed deep to the sternocleidomastoid (pp. 170; 172). The brachial plexus is formed by ventral rami from C5 to T1. Observe these **rami** passing laterally in the interval between the anterior and middle scalene muscles (Figure 9.6). To follow the brachial plexus into the axilla where the infraclavicular part has already been dissected, remove the central and medial thirds of the clavicle, including the sternoclavicular joint, with bone cutters and scalpel. While removing the clavicle, identify and remove the subclavius muscle deep to the clavicle. Do not disturb the loose fascia and structures deep to the subclavius at this time. The supraclavicular part of the brachial plexus in the posterior triangle and root of the neck includes the C5 to T1 ventral rami, superior, middle and inferior trunks and several branches of the rami and trunks. Identify the **fifth** and **sixth cervical rami,** which form the **superior trunk,** the **seventh ramus** which forms the **middle trunk,** and the **eighth cervical** and **first thoracic rami,** which form the **inferior trunk.** The inferior trunk is found deep to the subclavian artery and the superior and middle trunks are superior to this artery. Identify from the superior trunk the **suprascapular nerve,** which passes toward the scapular notch in the connective tissue at the superior scapula deep to the trapezius. The **dorsal scapular nerve** arises from the fifth ventral ramus to pass through the scalene muscles. The **long thoracic nerve** branches from the fifth, sixth and seventh rami and passes posterior to the brachial plexus

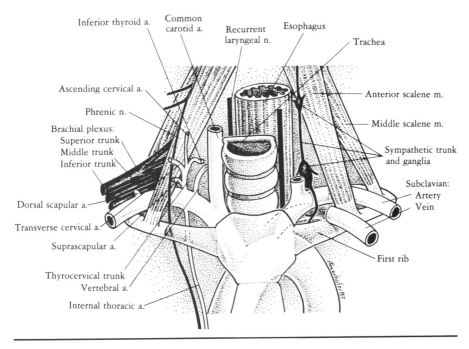

Figure 9.6

and subclavian artery to enter the axilla. Review its course in the axilla.

The last major dissection of the inferior part of the posterior triangle and root of the neck is that of the **subclavian artery** and **vein** (pp. 156–157; 170; 172). The subclavian artery system (Figure 9.6) is the primary arterial supply to the upper limb, shoulder and pectoral region. The right subclavian artery branches from the brachiocephalic trunk, whereas the left subclavian artery branches directly from the aortic arch. Both arteries enter the root of the neck posterior to the sternoclavicular joint. To dissect this arterial system, cut the first rib near its attachment to the manubrium sterni anterior to the subclavian artery. This maneuver allows elevation of the manubrium to observe the subclavian vessel deeply. Now, carefully clean away the fascia previously exposed deep to the subclavius to dissect the subclavian artery and vein. Preserve the arterial branches described in the text that follows during removal of this fascia. Especially watch for the suprascapular artery and nerve, which pass deep to the clavicle toward the superior border of the scapula.

On each side, the subclavian artery arches superiorly and laterally over the first rib, where its name changes to axillary artery as it enters the axilla. Most of this arched course is deep to the clavicle.

Observe that in its arched course in the root of the neck, the artery passes between the anterior and middle scalene muscles and is separated from the subclavian vein by the anterior scalene muscle (Figure 9.6). The position of the anterior scalene muscle typically divides the artery into three parts (p. 156): the first part, which is medial to the muscle, the second part, which is posterior to the muscle, and the third part, which is lateral to the muscle.

A. First Part of the Subclavian Artery

The first part of the subclavian artery has different origins on each side of the body, as noted previously. In the root of the neck, the internal jugular vein is anterior and the vagus and phrenic nerves are parallel to the artery on its anterior side. The first part of the artery usually grooves the lung and pleura on both sides. There are three branches from this segment of the subclavian artery: the vertebral artery, thyrocervical trunk and lateral thoracic artery.

1. Vertebral Artery. The **vertebral artery** arises from the subclavian artery at the level of the cervicothoracic sympathetic ganglion. Carefully clean the fascia between the subclavian artery and the inferior course of the common carotid artery to identify the deep vertebral artery. You may have to remove some of the veins found in this area to expose the artery. Note the relationships of the artery to the cervical sympathetic trunk and the middle (level of the sixth cervical vertebra) and inferior (level of the seventh cervical vertebra) cervical ganglia. The vertebral artery passes ventral to the inferior rami of the brachial plexus and enters the costotransverse foramen of the sixth cervical vertebra to ascend through the successive vertebral foramina until the artery finally reaches the cranial cavity through the foramen magnum. Except for spinal arteries, it provides no major branches in the neck but supplies the intracranial structures.

2. Thyrocervical Trunk. The **thyrocervical trunk** arises from the first part of the subclavian artery just distal to the vertebral artery. It is short and can be identified ascending the medial border of the anterior scalene muscle. From the trunk, identify the **inferior thyroid artery** at the medial border of the anterior scalene muscle. At the level of the cricoid cartilage, the inferior thyroid artery turns medially to pass posterior to the carotid sheath to supply the inferior pole of the thyroid and parathyroid glands. The next two branches of the thyrocervical trunk, the suprascapular and transverse cervical arteries, pass laterally across the ventral surface of the anterior scalene muscle and phrenic nerve. These two vessels hold the phrenic nerve to the scalene muscle and must be carefully separated from the prevertebral fascia. Identify the more inferior **suprascapular artery** as it passes across the anterior scalene muscle to course deep to the clavicle. The **suprascapular nerve** joins the artery deep to the clavicle. Follow the artery and nerve to the superior border of the scapula to reach the scapular notch, where its distribution on the dorsal surface of the scapula has been dissected in the posterior shoulder. The third branch is the **transverse cervical artery.** This artery is superior and parallel to the suprascapular artery and also crosses the anterior scalene muscle and phrenic nerve. The transverse cervical artery passes dorsally across the posterior triangle to course deep to the trapezius muscle. The dorsal scapular artery (discussed later) is often a branch of the transverse cervical artery.

3. Internal Thoracic Artery.   The third branch from the first part of the subclavian artery is the **internal thoracic artery.** This artery branches from the undersurface of the subclavian artery opposite the origin of the thyrocervical trunk. It descends vertically, posterior to the medial end of the clavicle and subclavian vein, to enter the superior mediastinum through the superior thoracic aperture. There are no branches in the neck, but the internal thoracic artery supplies the anterior chest wall and pericardium, as described in the discussion of the thorax (see Chapter 5).

B. Second Part of the Subclavian Artery

The second part of the subclavian artery is deep to the anterior scalene muscle. It usually has only one branch, the **costocervical trunk.** Being posterior to the anterior scalene muscle, this artery is better seen while looking superiorly from the thorax. The costocervical trunk passes over the apex of the lung and cervical fascia. It then divides into the posterior intercostal artery, which supplies the first and second posterior intercostal spaces, and the deep cervical artery, which supplies the deep muscles of the lateral and posterior neck.

C. Third Part of the Subclavian Artery

The last part of the subclavian artery is lateral to the anterior scalene muscle (p. 172). This part of the artery parallels the subclavian vein, which is ventral and inferior. The inferior trunk of the brachial plexus passes posterior to the third part of the subclavian artery and the middle and upper trunks are superior. The **dorsal scapular artery** is usually the only branch from the third part. However, its origin is highly variable, branching from the transverse cervical artery in about 30 percent of cases. If this artery is present, note that it weaves through the trunks of the brachial plexus to reach the deep surface of the levator scapulae, supplying this muscle and the rhomboid muscles.

Also, observe the position and relationships of the **subclavian vein** and note its only tributary, the external jugular vein. The **thoracic duct** enters the junction of the subclavian and internal jugular veins. Review the course of the thoracic duct from the thorax into the root of the neck.

78. SCALP

The scalp (p. 83) is a five-layer unit that covers the skull. From superficial to deep, the layers are the skin, a dense subcutaneous connective tissue, a muscular and aponeurotic layer (galea aponeurotica), loose areolar tissue and the periosteum (pericranium). The first three layers are bound together tightly and not easily separated. By contrast, the loose areolar tissue separates the muscular layer from the skull, allowing movement of the scalp. Blood vessels and nerves of the scalp arise from many sources and run in the dense subcutaneous layer. It is not necessary to dissect these layers in detail, but their relationships should be reviewed in your textbook. On the dry skull, identify the following structures: the **frontal, parietal, occipital** and **temporal** (squamous portion) **bones** and the **coronal, lambdoidal** and **sagittal sutures.** If the scalp has not previously been cut during removal of the calvaria and brain, make skin incisions to divide the scalp into four flaps. Blunt dissect the muscular layer from the other layers and note the **occipitalis** and **frontalis** muscles and the intervening galea aponeurotica. Pull the flaps of the scalp away from the skull and observe the periosteum that covers the skull. Scrape away this tissue to clean the calvaria.

79. CRANIAL CAVITY

If the calvaria has not been cut and the brain removed, follow instructions for these dissections given by the course personnel for their removal.

A. Bony Landmarks

Remove the calvaria and examine the inner surface of the cranial vault (pp. 34–36; 64). Identify the **anterior, middle** and **posterior cranial fossae** in the floor of the vault and outline their boundaries (Figure 9.7). Within each fossa, identify the positions of the **major**

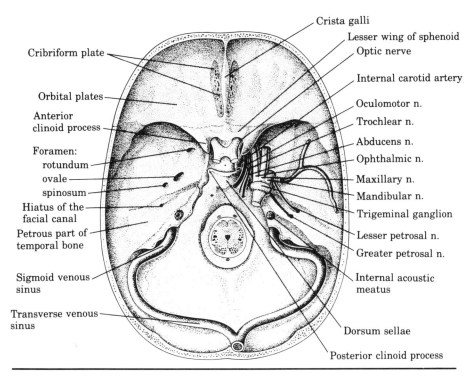

Figure 9.7

foramina and review their **contents** and the sites of exit of the cranial nerves. Compare these foramina with those of a dried skull. The foramina are not easily visible in the cadaver because the dura mater still lines the inner surface of the cranium (the arachnoid and pia mater were removed with the brain).

In the anterior cranial fossa of a dried skull, observe the **orbital plate** of the **frontal bone, cribriform plate of the ethmoid bone, crista galli** (upward extension of the ethmoid bone), **lesser wing of the sphenoid bone, anterior clinoid processes** and foramen cecum.

In the middle fossa, identify the **body of the sphenoid bone (sella turcica, dorsum sellae** and **posterior clinoid processes), greater wing of the sphenoid bone, squamous part of the temporal bone, carotid canal** and **groove, petrous part of the temporal bone, foramina spinosum, ovale** and **rotundum, superior orbital fissure** and hiatus of the facial canal.

In the posterior fossa, identify the **occipital bone,** posterior surface of the **petrous part of the temporal bone, dorsum sellae,** grooves for the **sigmoid** and **transverse venous sinuses, jugular foramen, foramen magnum, hypoglossal canal, internal auditory meatus** and **internal occipital protuberance.**

B. Meninges

The brain and brain stem are covered by the same three meningeal layers that surround the spinal cord. At the foramen magnum, the layers are continuous. The arachnoid and pia mater have the same relationships to the brain as they do to the spinal cord. These two layers were removed with the brain and are not present in the cranial vault. However, the cranial dura differs from the spinal dura. The cranial dura is composed of two layers that are usually fused: the periosteal dura and meningeal dura. The periosteal dura, the outer layer, is fused to the inner surface of the cranial bones and serves as their periosteum. It can be easily stripped from the bones. The true, or meningeal, dura mater is directly continuous with the spinal dura mater at the foramen magnum and surrounds the brain. The meningeal layer forms duplications that extend deeply between parts of the brain, providing support and stability. Identify and examine the following duplications of the dura: **falx cerebri, tentorium cerebelli, falx cerebelli** and **diaphragma sellae** within the cranium (pp. 83–85). Identify the attachments and the extent of these double folds of the dura.

C. Dural Venous Sinuses

In certain regions of the cranium, the periosteal and meningeal layers are separated to form endothelium-lined dural venous sinuses (pp. 83–85). These sinuses receive venous flow from the cerebral, cerebellar, meningeal, diploic and emissary veins and eventually drain into the internal jugular vein. Identify the following sinuses by opening the dural reflections and exposing their lumina:

1. The **superior sagittal** sinus is located in the superior midline attachment of the falx cerebri. Note the lateral extensions of this sinus (venous lacunae), which contain the arachnoid granulations that project into the sinus.

2. The **inferior sagittal sinus** can be observed in the inferior margin of the falx cerebri.

3. The **straight sinus** is located in the attached surface of the falx cerebri and tentorium cerebelli.

4. The **confluence of sinuses** is found at the junction of the straight and transverse sinuses at the internal occipital protuberance.

5. **Transverse sinuses** are identified on both sides in the poste-

rior arch of the tentorium cerebelli, where the tentorium attaches to the occipital bone.

6. **Sigmoid sinuses** groove the base and posterior surface of the petrous and mastoid parts of the temporal bones on each side to reach the jugular foramen.

7. **Cavernous sinuses** are found between the two layers of dura on either side of the body of the sphenoid bone. To open the cavernous sinus, carefully separate with a probe the layers of dura along the cut edge of the tentorium at the superior surface of the petrous bone. The cavernous sinus is unique because the **oculomotor, trochlear, ophthalmic** and **maxillary nerves** are found in its lateral wall. Carefully identify these nerves attached to the lateral wall as you separate the layers of dura, being especially careful not to destroy the extremely small trochlear nerve. Continue to strip the dura from the floor of the middle cranial fossa and note the relationship of the cavernous sinus to the **semilunar (trigeminal) ganglion,** which is located at the apex of the petrous part of the temporal bone (Figure 9.7). Identify this ganglion and the formation of the **ophthalmic, maxillary** and **mandibular nerves** from it (pp. 64; 69–71). Follow each of these nerves to its exit through its respective foramen. As the cavernous sinus is exposed by retracting the structures in the lateral wall, note that within the sinus are the **internal carotid artery** and **abducens nerve** (lateral to the artery). This artery and nerve are covered by endothelium.

Other, smaller sinuses include the occipital, basilar and superior and inferior petrosal sinuses. Review these and the major sinuses listed previously, noting their positions and routes of venous drainage from the cranial cavity into the internal jugular vein at the jugular foramen.

Be sure to identify all 12 cranial nerves in the cranial vault and recognize their points of exit from the cranial cavity. Also, study the intracranial courses and relationships of the **vertebral** and **internal carotid arteries.**

80. PHARYNX

The pharynx is a visceral structure of the neck that has respiratory and digestive functions. Posteriorly, it lies adjacent to the first six cervical vertebrae and occipital bone. Anteriorly, it opens into the nasal, oral and laryngeal cavities. It is described as a fibromuscular tube with walls that are deficient anteriorly. Superiorly, it is attached to the base of the skull. Inferiorly, it extends to the level of the cricoid cartilage, where it is continuous with the esophagus. Laterally, it is related to the carotid sheath and structures within the anterior cervical triangle. The pharynx is composed of three circular and three longitudinal muscles.

The pharynx is dissected from a posterior approach. The dissection involves an anterior separation, or disarticulation, of the head and structures of the neck from the vertebral column at the plane between the prevertebral and the buccopharyngeal fasciae (retropharyngeal space). The atlanto-occipital joint is disarticulated between the occipital condyles and the atlas. This dissection is carried out with the cadaver in the supine position.

A. Disarticulation of the Head

Step 1. If the suboccipital triangle has not been dissected, flex the head inferiorly toward the chest. Remove all the posterior skin and make a transverse cut to separate all the muscle layers on the posterior neck from the posterior surface of the occipital bone at the superior nuchal lines. Deep to these inferiorly reflected muscles, identify the posterior arch of the atlas and the space between the occipital bone and the atlas.

Step 2. A posterior wedge must be removed from the skull, as indicated in Figure 9.8. This wedge is created by making two saw cuts in a line *posterior* to the jugular foramen and *parallel* to the long axis of the petrous part of the temporal bone on each side of the skull. Stabilize the head and be sure that the two saw cuts intersect the *anterior* margin of the foramen magnum. After the cuts are made, remove the wedge by cutting any adherent muscle or skin.

Step 3. With the head flexed toward the chest, identify the atlanto-occipital joint and insert the chisel into the interval between the atlas and the occipital condyles (Figure 9.9). Use the mallet and chisel to break through the structures that form the atlanto-occipital joint on each side of the midline until the head is free from its articulation with the atlas. During this procedure, a part of the occipital clivus between the two condyles must also be chiseled. This part is indicated by a dotted line in Figure 9.9. To free the head and pull it forward from the vertebral column, cut the prevertebral muscles that still attach the head to the ventral surface of the vertebral column. Begin cutting the prevertebral and scalene muscles at each side of the cervical vertebrae in the floor of the posterior cervical triangle. These muscles can be cut after rocking the head forward after disarticulation and must be cut carefully to avoid damaging the entire cervical plexus in the posterior triangle of the neck. Keep the plexus

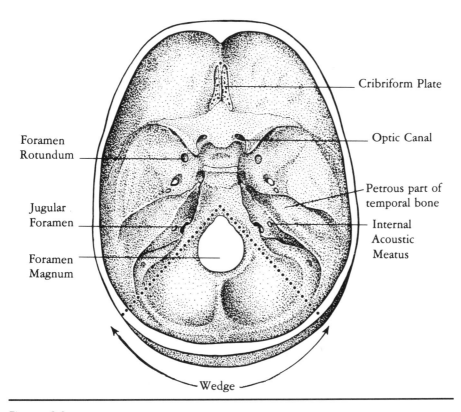

Cribriform Plate

Optic Canal

Foramen
Rotundum

Petrous part of
temporal bone

Jugular
Foramen

Internal
Acoustic
Meatus

Foramen
Magnum

Wedge

Figure 9.8

in sight as you pull the head forward. The first and second cervical
ventral rami usually have to be sacrificed as the muscles are cut at
each side of the disarticulated head, but C3 and C4 are left intact. Do
not disrupt any vascular structure already dissected in the neck. The
head should now be pulled forward from the vertebral column. The
separation is between the cervical prevertebral and buccopharyngeal
fasciae. The space thus developed is the **retropharyngeal space,**
which leads through the superior thoracic aperture to the mediastinum of the thorax below. Be sure to cut any prevertebral muscles
that remain attached to the skull.

B. Exterior of the Pharynx

After the head and structures of the neck have been pulled forward,
carefully remove the **buccopharyngeal fascia** from the central,
posterior surface of the pharynx (Figure 9.10). Identify the circular
superior, middle and **inferior pharyngeal constrictor** muscles
(pp. 152–155), noting how each muscle overlaps the adjacent muscle

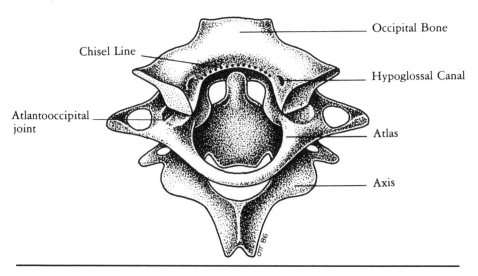

Figure 9.9

above. The separation between these muscles may not be obvious. Review the relationships of the middle and inferior constrictor muscles to the hyoid bone and thyroid and cricoid cartilages. Review the origins of these muscles and observe their posterior insertions into the **pharyngeal raphe** (p. 153). The raphe is attached to the **pharyngeal tubercle** of the occipital bone (p. 28) superiorly and is continuous with the esophagus inferiorly. The muscles are deficient anteriorly, where the pharynx communicates with the nasal, oral and laryngeal regions. There is an interval, or gap, between the superior constrictor muscle and the base of the skull, which is filled with the dense pharyngobasilar fascia. This fascia is the submucosal layer of the pharynx deep to the muscles and is penetrated by the auditory (eustachian) tube and levator veli palatini muscle. The second gap that appears between the superior and the middle constrictor muscles conducts the stylopharyngeus muscle and glossopharyngeal nerve (discussed later). The three longitudinal pharyngeal muscles will be studied later.

Complete the removal of the buccopharyngeal fascia on the lateral aspect of the posterior pharynx and carefully dissect the courses (Figure 9.10) of the **internal carotid artery, internal jugular vein, last four cranial nerves** and **sympathetic trunk** (p. 155). The **internal carotid artery** ascends within the carotid sheath to the carotid canal at the base of the skull, where the sheath invests the glossopharyngeal, vagus, accessory and hypoglossal nerves. Elevate the sternocleidomastoid superiorly and review the course of the **acces-**

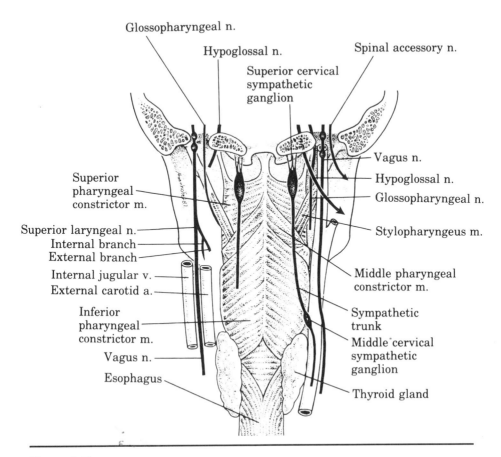

Glossopharyngeal n.

Hypoglossal n.

Spinal accessory n.

Superior cervical
sympathetic
ganglion

Superior
pharyngeal
constrictor m.

Superior laryngeal n.
Internal branch
External branch

Internal jugular v.
External carotid a.

Inferior
pharyngeal
constrictor m.

Vagus n.

Esophagus

Vagus n.

Hypoglossal n.

Glossopharyngeal n.

Stylopharyngeus m.

Middle pharyngeal
constrictor m.

Sympathetic
trunk

Middle cervical
sympathetic
ganglion

Thyroid gland

Figure 9.10

sory nerve, which enters the sternocleidomastoid after passing lateral to the **internal jugular vein.** The **vagus nerve** can be identified easily by its position between the internal carotid artery and the internal jugular vein in the carotid sheath. Look for the two sensory ganglia on the vagus nerve at the base of the skull. Identify the origin of the **superior laryngeal nerve** from the vagus nerve posteriorly on the pharynx. Follow this nerve posterior to the carotid sheath, where its **internal** and **external branches** can be observed (Figure 9.11). The **hypoglossal nerve** exits the hypoglossal canal posterior to the jugular foramen. The hypoglossal nerve then courses lateral to the internal carotid artery, looping the occipital artery to enter the carotid triangle, where it has previously been dissected. Note this relationship. Lateral to the pharynx at the level of the atlas, the hypoglossal nerve receives nerve fibers from the first and second cervical spinal ventral rami. These spinal nerve fibers course

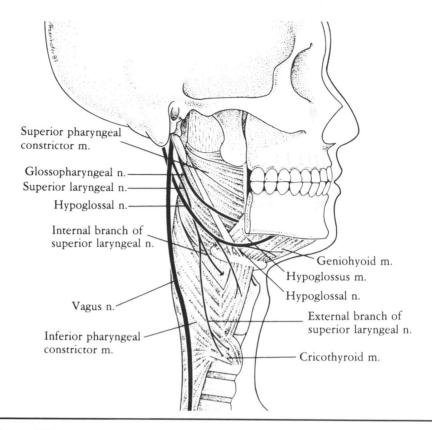

Superior pharyngeal constrictor m.

Glossopharyngeal n.

Superior laryngeal n.

Hypoglossal n.

Internal branch of superior laryngeal n.

Geniohyoid m.

Hypoglossus m.

Hypoglossal n.

Vagus n.

External branch of superior laryngeal n.

Inferior pharyngeal constrictor m.

Cricothyroid m.

Figure 9.11

with the hypoglossal nerve a short distance and leave it in the carotid triangle to form the **superior ramus of the ansa cervicalis.** The **glossopharyngeal nerve** can be identified more laterally and superior to the hyoid bone by retracting the internal carotid artery to expose the **stylopharyngeus** muscle (p. 153). The stylopharyngeus is one of the three longitudinal muscles and enters the pharynx between the superior and the middle constrictor muscles at the level of the greater cornu of the hyoid bone. Carefully dissect the fascia from the posterior surface of the stylopharyngeus to identify the **glossopharyngeal nerve** as it passes posterior and lateral to the muscle before piercing the pharyngeal wall between the superior and the middle constrictor muscles. Two sensory ganglia are present on the glossopharyngeal nerve at the jugular foramen.

Identify the **sympathetic trunk,** which descends posterior to the carotid artery in the prevertebral cervical fascia from the base of the skull into the thorax. Note that there are typically three **chain ganglia** along the sympathetic trunk in the neck. The **superior cervical**

ganglion (p. 172) is the largest and is located at the level of the second cervical vertebra. The **middle ganglion** is variable in size and is found at or below the level of the cricoid cartilage. The **inferior ganglion** is often fused with the first thoracic ganglion (cervicothoracic or stellate ganglion) and lies at the level of the seventh cervical vertebra. Representative branches of the ganglia (gray rami communicantes, pharyngeal and cardiac nerves) may be identified. From the vagus and glossopharyngeal nerves, find examples of their contributions to the pharyngeal plexus on the middle pharyngeal constrictor (p. 155). Review in your textbook the function of each of these three nerves (vagus, glossopharyngeal and sympathetic) in the innervation of the pharynx. **Cardiac branches** of the vagus nerve and sympathetic trunk arise posterior to the pharynx. Also, identify the **ascending pharyngeal artery,** which branches from the posterior surface of the external carotid artery and ascends on the lateral surface of the pharynx.

C. Interior of the Pharynx

Internally, the pharynx is divided into the **nasal pharynx** between the base of the skull and the soft palate, the **oral pharynx** between the soft palate and the base of the tongue and the hyoid bone, and the **laryngeal pharynx** between the hyoid bone and the level of the cricoid cartilage. These parts of the pharynx open anteriorly into the nasal, oral and laryngeal cavities, respectively. The nasal pharynx and most of the oral pharynx will be studied later after a sagittal section of the head is made.

To view the inner parts of the laryngeal pharynx, cut through the pharyngeal raphe from the base of the skull to the junction of the pharynx and esophagus (pp. 150–151). Open the posterior pharyngeal wall and note the **laryngeal inlet** into the vestibule of the larynx. Identify the superior part of the **epiglottis** and the **aryepiglottic folds** that border this inlet and the **piriform recess** lateral to it. Identify the dorsum and base of the tongue, which are connected to the epiglottis by the **median** and **lateral glossoepiglottic folds.**

81. LARYNX

The larynx occupies the visceral area of the neck with the pharynx and both structures are enclosed by the pretrachael and buccopharyngeal fasciae. The larynx is the air passage between the pharynx and trachea and is uniquely designed for the production of sound. It is composed of cartilages, membranes, ligaments and muscles that alter the shape and position of the laryngeal structures to produce sound and control movement of air. Note the location of the larynx anterior to the fourth, fifth and sixth cervical vertebrae (p. 147). Anterior and lateral to the larynx are the thyroid gland, infrahyoid muscles and common carotid arteries.

In your atlas and textbook, review the shapes, positions and parts of the **thyroid, cricoid, arytenoid** and **epiglottic** cartilages, which form the skeleton of the larynx (pp. 146–147). Describe the articulation between the thyroid and the cricoid cartilages (**cricothyroid joint**) and between the cricoid and the arytenoid cartilages (**cricoarytenoid joint**). Define the movements at these joints.

On the dorsal surface of the larynx, separate the aryepiglottic folds that border the laryngeal inlet and insert your finger into the vestibule. Palpate the arytenoid cartilage at the posterior part of the inlet and identify the **vestibular** and **vocal folds** and **rima glottidis** (p. 149) while looking in the vestibule. The vestibular folds are lateral to the vocal folds.

At the posterior pharyngeal surface of the larynx, strip the mucosa from the entire piriform recess and posterior surface of the lamina of the cricoid cartilage. Deep to the mucosa of the piriform recess, two branches of the vagus nerve distribute to the larynx. The **inferior laryngeal nerve** courses posterior to the cricothyroid joint in the lower part of the recess (p. 150). The inferior laryngeal nerve is the superior continuation of the recurrent laryngeal nerve superior to the cricothyroid joint. The inferior laryngeal nerve is sensory to the laryngeal mucosa inferior to the vocal folds and innervates all the intrinsic muscles of the larynx except the cricothyroid muscle, which is innervated by the external branch of the superior laryngeal nerve. Superiorly, the **internal branch of the superior laryngeal nerve** is located deep to the mucosa at the thyrohyoid membrane. Identify these fibers in the mucosa superior and lateral to the arytenoid cartilage. The internal branch is the principal sensory nerve of the larynx superior to the vocal folds. It also supplies parasympathetic fibers to glands at the base of the tongue, aryepiglottic fold and interior of the larynx and supplies the taste buds at the base of the tongue.

Deep to the mucosa that covers the lamina of the cricoid cartilage, identify the position of the arytenoid cartilage and note its relationship to the aryepiglottic fold; the **posterior cricoarytenoid** muscle, which is attached to the posterior lamina of the cricoid cartilage and the muscular process of the arytenoid cartilage; and the **transverse** and **oblique arytenoid** muscles, which are attached to the arytenoid cartilage (p. 148). Note that the fibers of the oblique arytenoid muscle are extremely small and sweep laterally, entering the aryepiglottic fold to reach the epiglottis as the **aryepiglottic** muscle.

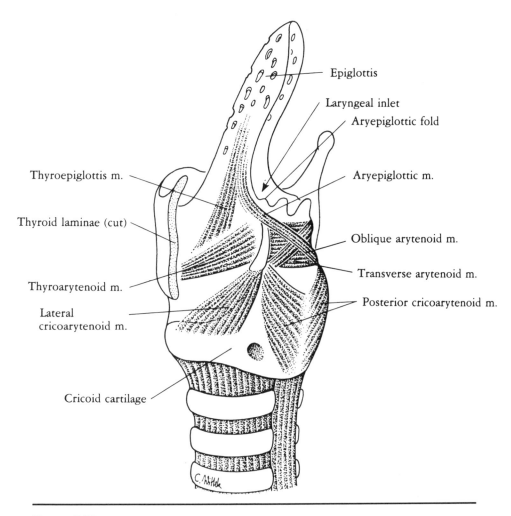

Thyroepiglottis m.

Thyroid laminae (cut)

Thyroarytenoid m.

Lateral
cricoarytenoid m.

Cricoid cartilage

Epiglottis

Laryngeal inlet

Aryepiglottic fold

Aryepiglottic m.

Oblique arytenoid m.

Transverse arytenoid m.

Posterior cricoarytenoid m.

Figure 9.12

The deep dissection of the larynx is from the anterolateral aspect, where one lamina of the thyroid cartilage will be removed on one side (Figure 9.12). On the ventral surface of the larynx, identify the **thyrohyoid** muscle and its **nerve to the thyrohyoid.** These fibers travel initially with the hypoglossal nerve and can be seen leaving it to enter the thyrohyoid. Reflect the thyrohyoid and note the **thyrohyoid membrane** deep to the muscle. Trace the courses of the **internal branch of the superior laryngeal nerve** and **superior laryngeal artery** through the membrane into the larynx. Identify the **cricothyroid** muscle on the ventral surface of the larynx, extending between the cricoid and thyroid cartilages. Note the direction of its fibers. Look for the small **external branch of the su-**

perior laryngeal nerve, which travels with the superior thyroid artery to reach the cricothyroid.

To remove the lamina of the thyroid cartilage, disarticulate the cricothyroid joint on this side. Along the superior, posterior and inferior margins of the thyroid lamina, cut and reflect all the remaining fascial and muscular (cricothyroid and inferior pharyngeal constrictor) attachments. Lift the posterior edge of the lamina and pull it anteriorly and medially while carefully separating the deeper muscles and fascia from the inner surface of the lamina as it is reflected. With this anterolateral view, identify the **lateral cricoarytenoid, thyroarytenoid** and **thyroepiglottic** muscles (p. 148). Although not visible in the dissection, the vocalis muscle is located in the vocal fold and is usually considered to comprise the most medial segments of the thyroarytenoid muscle. Review the origins, insertions, actions and innervations of all the laryngeal muscles.

To examine the internal aspects of the larynx, cut the *dorsal* midline from the laryngeal inlet inferiorly to the upper rings of the trachea (p. 149). With the larynx opened, review the positions of the **vestibular (false)** and **vocal (true)** folds. The cavity of the larynx superior to the vestibular folds is the **vestibule.** The walls of the vestibule deep to the mucosa are formed by the quadrangular membrane. The inferior free border of this membrane forms the vestibular ligament. Between the vestibular and vocal folds is the **ventricle** of the larynx, which leads to a lateral recess, the laryngeal saccule. The ventricular saccule contains mucous glands that moisten the vocal folds. Inferior to the vocal folds is the space known as the **infraglottic cavity,** the walls of which are formed by the conus elasticus (cricothyroid ligament) deep to the mucosa. The thickened, superior free edge of the conus forms the vocal ligament, which extends between the thyroid cartilage anteriorly and the vocal process of the arytenoid cartilage posteriorly. Review the movements of the vocal folds.

Inferior to the cricoid cartilage, the larynx continues as the trachea. Note the position of the trachea in the neck and identify the esophagus lying between the trachea and the vertebral column. Dissect the courses of the recurrent laryngeal nerves between the esophagus and the trachea as they pass deep to the lateral lobes of the thyroid gland.

82. SUBMANDIBULAR REGION

The submandibular, or suprahyoid, region is the superior division of the anterior cervical triangle and is bounded by the **anterior** and **posterior** bellies of the **digastric** muscles and inferior border of the

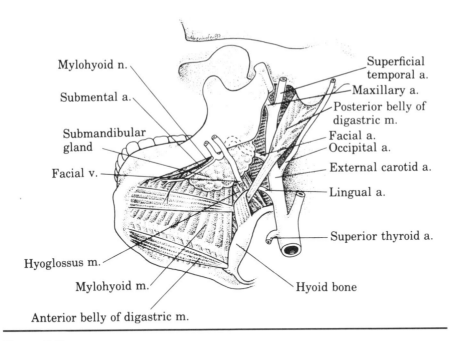

Figure 9.13

mandible. This region was not dissected with the other triangles of the neck because most of the structures within the submandibular triangle are closely related to the face and floor of the oral cavity.

The superficial layer of the cervical fascia creates a roof for the submandibular triangle. Superficial structures are the **great auricular nerve,** branches of the **transverse cervical nerve, submental vein** and the **marginal mandibular** and **cervical branches of the facial nerve.** The **platysma** muscle should be removed from its attachment to the mandible.

Basic to an understanding of the relationships of structures within the submandibular triangle (Figure 9.13) is appreciation of the orientation of the muscles in the triangle. To identify these muscles, free and elevate the submandibular gland from its fascia. This **superficial portion of the submandibular gland** is covered by a loose sheath derived from the superficial layer of the cervical fascia. Dissect and observe the courses of the facial vein superficial to the gland and the facial artery, which is deep to it.

Mention has already been made of the anterior and posterior bellies of the digastric muscle (p. 165). Review the positions, attachments and relationships of these bellies. In the submandibular triangle, identify the **mylohyoid** muscles, which extend between the two mandibles deep to the anterior belly of the digastric muscle. The mylohyoid muscle arises from the mylohyoid line of the mandible

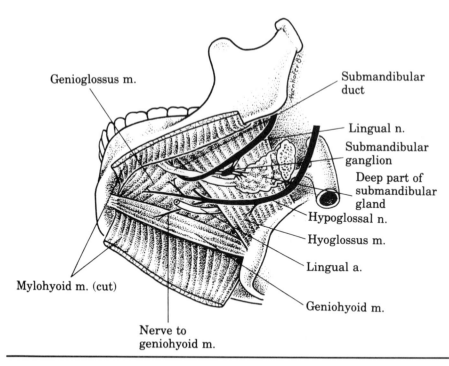

Genioglossus m.

Submandibular duct

Lingual n.

Submandibular ganglion

Deep part of submandibular gland

Hypoglossal n.

Hyoglossus m.

Lingual a.

Mylohyoid m. (cut)

Geniohyoid m.

Nerve to geniohyoid m.

Figure 9.14

and slopes inferiorly and medially to insert on the **mylohyoid raphe** and **hyoid bone** (p. 139). Superficial to the mylohyoid muscle, observe the **submental vessels** and **mylohyoid nerve,** which innervates the mylohyoid and anterior belly of the digastric muscle (p. 139). Trace the origin of the submental artery from the facial artery.

Parallel to the posterior belly of the digastric muscle, identify the **stylohyoid** muscle (p. 138). This muscle splits to pass around the intermediate tendon of the digastric muscle and inserts on the hyoid bone. The stylohyoid and posterior belly of the digastric muscle are innervated by branches of the facial nerve.

The other muscles in the submandibular triangle are deep to the mylohyoid muscle (Figure 9.14). To reflect this muscle, first cut the attachment of the anterior belly of the digastric muscle from the mandible and reflect it inferiorly. Then, cut the fibers of the mylohyoid along the raphe between the mandible and the hyoid bone and from the hyoid bone. Reflect the mylohyoid laterally. Identify the **geniohyoid, genioglossus** and **hyoglossus** muscles deep to the mylohyoid (p. 139). Review the attachments and innervations of these muscles. The stylohyoid, mylohyoid, geniohyoid and digastric muscles are collectively referred to as the **suprahyoid** muscles. They elevate the tongue and floor of the mouth during swallowing, elevate the larynx and assist in depressing the mandible during opening of the mouth. The **hyoglossus** is an extrinsic muscle of the tongue and a key landmark in the submandibular region, as described in the text that follows.

Deep to the mylohyoid, between it and the hyoglossus, identify the **deep part of the submandibular gland, sublingual gland, submandibular duct, lingual nerve, submandibular ganglion** and **hypoglossal nerve** (p. 139). Follow the submandibular duct as it leaves the deep part of the submandibular gland and passes superficial to the hyoglossus and then superficial to the genioglossus to reach the floor of the mouth at the sublingual caruncle deep to the tongue (Figure 9.14). A finger-like band of glandular tissue is often parallel to and covers the duct. Along the distal end of the duct, locate the **sublingual gland** deep to the mylohyoid (p. 140). This elongated gland lies just deep to the sublingual mucosa on the floor of the oral cavity. Identify the **lingual nerve** (a branch of the mandibular division of the trigeminal nerve) as it enters the submandibular region posteriorly from the infratemporal fossa by passing close to the medial surface of the mandible. In the submandibular triangle, the lingual nerve curves anteriorly by the deep part of the submandibular gland and then passes lateral, inferior and medial to the submandibular duct before reaching the mucosa of the floor of the mouth and anterior two-thirds of the tongue. The lingual nerve provides general sensation to both of these areas. As the lingual nerve curves by the submandibular gland, identify the **submandibular ganglion,** which is suspended from the nerve. This ganglion is often embedded in the deep part of the submandibular gland. The ganglion is one of the four parasympathetic terminal ganglia of the head and is functionally related to the facial nerve, containing postganglionic cell bodies of neurons that provide secretory fibers to the submandibular and sublingual glands. Locate the **hypoglossal nerve** as it crosses the superficial surface of the hyoglossus and passes superiorly to innervate the intrinsic and extrinsic muscles of the tongue. The **lingual artery** enters the submandibular region from the carotid triangle and courses deep to the hyoglossus. Cut and reflect the hyoglossus transversely on one side to observe this deep course of the lingual artery. It supplies blood to the tongue and floor of the mouth by way of its dorsal and deep lingual branches, which arise deep to the hyoglossus.

Review the relationships and completely dissect the courses of the structures that pass deep to the posterior belly of the digastric muscle between the carotid and the submandibular triangles: the **facial artery, lingual artery** and **hypoglossal nerve.**

83. FACE

It is important to precede all the following dissections of the face and deeper areas of the head with a study of the bones of the skull. Your study of the skull should involve identification of the bones that comprise the various surfaces (lateral, anterior and base) of the skull and their main features, foramina and processes (pp. 25–27; 50). Review the internal features of the cranial cavity, which were studied previously.

On the anterior surface of the skull, identify the major bones and landmarks: the **frontal bone** (and **zygomatic process**), **nasal bone, lacrimal bone, maxilla** (and **frontal process**), **zygomatic bone** (and **frontal process**), **mandible** and the **supraorbital, infraorbital, mental** and **zygomaticofacial foramina.** Also, review the general areas of cutaneous innervation provided by branches of the ophthalmic, maxillary and mandibular divisions of the trigeminal nerve. Recall that the cutaneous branches of the cervical plexus (great auricular and lesser occipital) overlap the trigeminal cutaneous innervation at the lateral side of the face and scalp, lower border of the mandible and external ear.

A. Muscles of Facial Expression

The skin is to be incised and removed as illustrated in Figure 9.15. A vertical midline cut should be made from the frontal area of the scalp to the mental protuberance of the mandible. Encircle the base of the nose and make cuts around the margins of the orbit and the lips as indicated. The inferior end of this vertical incision at the mental protuberance will meet the incision that was previously made at the inferior border of the mandible during the dissection of the neck. Next, make a vertical cut superiorly to the scalp anterior to the ear. Reflect the skin medially. During reflection of the skin toward the midline area of the face, note the unique location of the muscles of facial expression within the subcutaneous fascia. Because the skin is extremely thin and tightly attached to the fascia and muscles, its removal requires careful attention.

The major **muscles of facial expression** are arranged at the orbit, nose and mouth (Figure 9.16). These muscles insert into the skin and are important in movements of the eyelids, nasal entrance and lips and cheeks. All these nerves are innervated by branches of the facial nerve and are thin and delicate, making dissection difficult. By observing several cadavers, you can observe most of the following muscles at the orbit, nose and mouth in the laboratory (pp. 58–59):

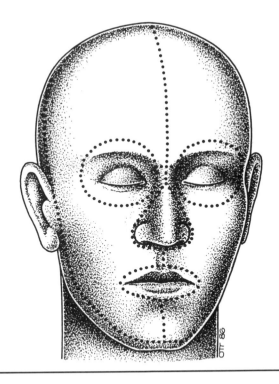

Figure 9.15

1. **Orbit.** Dissect the **orbicularis oculi** muscle, noting its wide expansion of fibers around the orbit. The outer fibers comprise the **orbital portion,** which functions in tight closure of the eyes. The smaller palpebral portion in the fascia of the eyelids is involved in blinking. Note the fibers of the **frontalis** muscle superior to the orbit.

2. **Nose and ears.** These muscles are minor and it is not necessary to dissect them.

3. **Mouth.** The muscles that surround the mouth are important not only because they are involved in facial expression, but also because they are vital in speech and chewing. Use blunt dissection with scissors and forceps to identify the following facial muscles: **levator labii superioris alaeque nasi** and **levator labii superioris,** which insert in the upper lip; **levator anguli oris, zygomaticus major** and **depressor anguli oris,** which insert in the angle of the mouth; and **depressor labii inferioris** and **mentalis,** which insert in the lower lip. Most of these muscles insert into the lips and collectively form the **orbicularis oris,** which serves as a sphincter of the mouth and cheek. Another muscle associated with the mouth is the **buccinator.** Identify this muscle in the deeper part of the cheek at

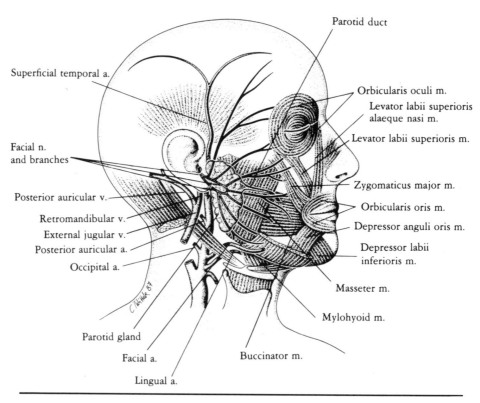

Parotid duct

Superficial temporal a.

Orbicularis oculi m.

Levator labii superioris alaeque nasi m.

Levator labii superioris m.

Facial n. and branches

Zygomaticus major m.

Orbicularis oris m.

Posterior auricular v.

Depressor anguli oris m.

Retromandibular v.

External jugular v.

Posterior auricular a.

Depressor labii inferioris m.

Occipital a.

Masseter m.

Mylohyoid m.

Parotid gland

Facial a.

Buccinator m.

Lingual a.

Figure 9.16

the anterior border of the masseter muscle. It can be seen approaching the angle of the mouth. The superior and inferior fibers of the buccinator decussate at the angle of the mouth to enter the upper and lower lips, thus contributing to the fibers of the orbicularis oris. Note the **buccal fat pad** superficial to the buccinator. Watch for the **buccal nerve,** a sensory branch of the mandibular nerve (C.N.V.) that is superior to the fat pad and deep to the masseter muscle.

B. Facial Nerve and Vessels

Carefully clean and identify the delicate, small branches of the **facial nerve** (Figure 9.16) on the deep surfaces of the facial muscles (pp. 74–76). They lie in the subcutaneous fascia and approach the muscles from the lateral (parotid) aspect of the face. In a later dissection, these nerves will be useful guides back to the main trunk of the facial nerve within the parotid gland.

From the submandibular (suprahyoid) region, follow onto the face the courses of the **facial artery** and **vein** (pp. 74–76). Note their relationships to the superficial part of the submandibular gland. The facial artery crosses the mandible at the anterior border of the masseter to enter the face (Figure 9.16). The course of the artery on the face is irregular and tortuous, passing the angle of the mouth and nose to terminate as the angular artery at the orbit. Name the major branches of the facial artery on the face. Observe the facial vein lying on the lateral side of the artery. It is a continuation of the angular vein at the orbit. On the face, the facial vein communicates with the deep pterygoid venous plexus and transverse facial vein. It is superficial in the submandibular triangle, where it usually terminates in the internal jugular vein.

The final step in this dissection is to locate examples of the major cutaneous nerves on the face. Isolate and locate the **supraorbital, infraorbital** and **mental nerves,** which exit the same-named **foramina** deep to the facial muscles. Observe the positions and relationships of the cutaneous nerves to the facial muscles and note their areas of sensory distribution.

84. PAROTID REGION

Before beginning this dissection, review the major bony landmarks on the lateral side of the skull. Identify the **frontal, parietal, squamous part of the temporal, greater wing of the sphenoid, maxilla** and **zygomatic bones, zygomatic arch, mandibular fossa, articular tubercle** and **external acoustic meatus** (pp. 25–26).

A. Bisection of the Head

The first step in your dissection will be to divide the head into two halves by bisecting it at the midline. This approach will allow the dissection team on each side of the cadaver better access to the head for dissection. This bisection also makes it possible to study some of the deep areas (such as the oral and nasal cavities) not accessible in an undivided head. Use the saw for cutting bony structures and the scalpel for soft structures. Begin the saw cut at the superior aspect of the skull on a line that begins at the midpoint of the frontal bone and passes posteriorly just lateral to the crista galli through the middle of the sphenoid and occipital bones and foramen magnum. Steady the head and cut just to one side of the nasal septum and through the midline of the hard palate. Cut the tongue and midline structures in the floor of the mouth with the scalpel. The midline cut through the mandible and hyoid bone must be made with the saw. With the scalpel, continue this midline cut to the *upper rings* of the trachea to meet the previous midline cuts in the posterior larynx. The two sides of the head is now separated for further dissection.

B. Parotid Gland

The parotid region occupies the lateral side of the face anterior to the ear (Figure 9.16). Describe the boundaries of this region. Deeply, the parotid gland extends into the retromandibular region to the lateral wall of the pharynx.

The **parotid gland** (pp. 74–75) fills the parotid region. The gland has a wide, triangular, superficial part that spreads over a large area of the lateral side of the face anterior to the ear. Deep to its superficial part, the gland narrows to a wedge-shaped structure that fills the retromandibular region. This narrow part (isthmus) of the gland is lodged between the posterior belly of the digastric muscle and styloid muscles posteriorly and medially and the ramus of the mandible and its attached masseter and medial pterygoid muscles anteriorly.

Observe that the superficial part of the gland is bounded by the zygomatic arch superiorly and by the ear, sternocleidomastoid and posterior belly of the digastric muscle posteriorly and overlies the masseter anteriorly (Figure 9.16) (p. 75). The **masseter** is one of the four muscles of mastication. It is quadrilateral in shape. Clean completely the lateral, anterior and posterior borders of the masseter and note its attachment between the zygomatic arch and the lateral surface of the angle of the mandible. Identify the **parotid duct,** which extends from the anterior border of the gland and passes across the face 1 cm below the zygomatic arch (p. 75). The duct crosses the masseter muscle, then penetrates deeply at the buccal fat pad through the buccinator muscle and opens into the vestibule of the oral cavity opposite the upper second molar tooth.

Most of the dissection of this region involves removal of the parotid gland. It is difficult to remove this gland because it is embedded in the dense parotid fascia, which is a superior continuation of the superficial layer of the cervical fascia. The parotid fascia covers the superficial and deep parts of the gland. Review in your textbook the extent of this fascial covering and its deeper relationships.

The gland must be removed carefully because most of the other structures in this region are intimately related to and embedded in the gland. The deepest structures are the **external carotid artery** and its last three branches. Superficial to these branches is the **retromandibular vein** and most superficial in the gland are the branches of the **facial nerve.** With forceps and scissors, begin removing the gland in small pieces, as described in the text that follows. The gland has to be removed from both sides of the head to permit the deeper dissections of the head.

As was indicated previously, the most superficial structures to be dissected free from the gland are the **facial nerve** and its branches (pp. 75–76). The branches and trunk of the facial nerve can be located by following back to the gland the previously identified peripheral branches of the facial nerve, which supply the facial muscles. These distal branches of the facial nerve are superficial to the masseter and fan from the anterior border of the parotid gland as **temporal, zygomatic, buccal, marginal mandibular** and **cervical branches** to reach the facial muscles (Figure 9.16). Identify these nerves in the fascia between the facial muscle and the anterior border of the parotid gland. Note that the large buccal branches run with the parotid duct.

At the anterior border of the gland, use scissors to tunnel deeply through the gland along each of the groups of facial nerve branches identified earlier. Cut away all the glandular tissue around and superficial to these nerves. In the deeper part of the gland, follow the nerves to the point where they form **temporofacial** and **cervicofacial** divisions, which may loop around the isthmus of the gland (p. 76). Cut away *all* the glandular tissue superficial to these divisions. From this point, the trunk of the facial nerve can be identified by following the fibers of the two facial divisions deeply toward the stylomastoid foramen. At this foramen, the facial nerve also provides fibers that innervate the auricular and occipitalis muscles of facial expression and the posterior belly of the digastric and stylohoid muscles.

With the superficial part and isthmus of the gland removed, identify the **retromandibular vein** deep to the nerve (Figure 9.16). Continue to remove all the glandular tissue from the surfaces of the vein. This vein is formed by the **superficial temporal** and **maxillary veins** in the superior part of the gland (pp. 75–76). Inferiorly, the retromandibular vein and **posterior auricular vein** join to form the **external jugular vein.** Dissections of these venous relationships can be facilitated by tunneling superiorly along the external jugular vein from the neck into the retromandibular area. It is important to clean the anterior border of the sternocleidomastoid superiorly to the mastoid process by removing any skin and fascia to establish continuity of the structures that pass between the lateral neck and the retromandibular region deep to the posterior belly of the digastric muscle. Remove any remaining fibers of the platysma.

With the retromandibular vein dissected, expose the **external carotid artery** deeply within the gland (pp. 76–78). Follow the superior course of this artery from the carotid triangle as it passes deep to the posterior belly of the digastric muscle to enter the retromandibular region on the deep surface of the parotid gland (Figure 9.16). Complete the removal of any remaining parotid tissue. Posterior to the mandible, the external carotid artery has three branches:

1. **Posterior auricular artery.** This artery arises from the posterior surface of the carotid artery at the superior border of the posterior digastric muscle and passes superiorly and laterally to the scalp posterior to the ear.

2. **Superficial temporal artery.** This artery is one of two terminal branches of the external carotid artery. Beginning posterior to the neck of the mandible, it courses superiorly across the posterior end of the zygomatic arch to reach the lateral surface of the scalp anterior to the ear (temporal area). Identify the **transverse facial artery,** a branch of the superficial temporal artery that arises in the superior part of the parotid gland. This branch passes across the face between the parotid duct and the zygomatic arch.

3. **Maxillary artery.** This artery is the second terminal branch and arises posterior to the mandibular neck. It courses medially and deep to the neck, reaching the infratemporal fossa, where it will be dissected.

In the dense fascia of the temporal region adjacent to the ear, the **auriculotemporal nerve** may be found posterior to the superficial temporal artery. This nerve is sensory to the scalp, parotid capsule and temporomandibular joint and branches from the mandibular nerve in the infratemporal fossa. It passes posterior to the mandibular neck to enter the parotid gland, where it crosses the zygomatic arch to reach the scalp.

With all these structures exposed, the isthmus and deep aspects of the parotid gland should have been removed. Review the relationships of these parts of the gland to the styloid process and its attached muscles posteriorly, mandible and its attached muscles anteriorly and lateral pharyngeal wall and carotid sheath deeply.

Now that the retromandibular region has been dissected, review and identify the three muscles that arise from the styloid process: the **stylohyoid, styloglossus** and **stylopharyngeus.** Describe their insertions and innervations. Note that the parotid region must be *completely* dissected before the next dissection is begun. The posterior surface of the masseter must be cleaned and exposed.

85. TEMPORAL AND INFRATEMPORAL FOSSAE

Beginning with this dissection, the deep spaces of the head will be studied. For each area, identify the major boundaries, structures,

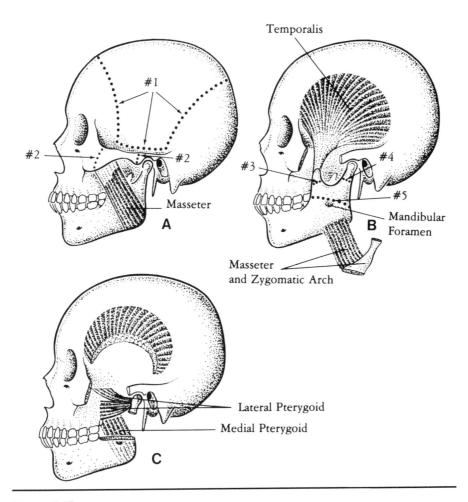

Figure 9.17

relationships and foramina and their contents. Be aware of the relationships of each deep space to other fossae and of the communications between the various foramina and fissures.

Begin the dissection of the temporal and infratemporal fossae by defining their boundaries, bony landmarks and foramina in your textbook and locating them on the dried skull. This dissection will expose the masticatory muscles, mandibular nerve and maxillary vessels.

The sequence and accuracy of this dissection are critical and should be followed exactly on both sides of the head (Figure 9.17). This work cannot begin until all the dissections of the parotid and retromandibular regions have been completed. Carefully follow the instructions given, which are illustrated in Figure 9.17.

Step 1. First, identify and palpate the **zygomatic arch** and its components. Superior to the zygomatic arch, the temporal fossa contains the temporalis muscle and its nerves and vessels. This fossa is covered by the temporal fascia and overlying galea aponeurotica of the scalp. These structures form a dense covering for the temporalis and have firm attachments to the superior border of the zygomatic arch. Expose these fascial layers on the lateral surface of the skull by removing any remaining skin from the temporal fascia and scalp. Define the attachments of these fascial layers to the bony margins of the temporal fossa and cut them away from the superior border of the zygomatic arch inferiorly and the bones anteriorly and posteriorly (Figure 9.17, A, Cut #1). Peel these fascial layers superiorly and identify the **temporalis** deep to them. The temporalis is another muscle of mastication and arises as a wide, fan-shaped structure from the temporal fascia and bony floor of the temporal fossa (p. 55). Note the courses and directions of the anterior and posterior fibers. Describe their functions. The fibers narrow and pass deep to the zygomatic arch to insert on all sides of the coronoid process of the mandible. This insertion is extensive. As will be seen later, some fibers continue down the anterior border of the mandible as far as the last molar tooth and have important relationships to the buccal nerve.

Step 2. The next step in dissecting the infratemporal fossa is to remove the zygomatic arch and attached fibers of the masseter (p. 55). Observe and clean completely the anterior and posterior borders of the **masseter** on the lateral surface of the mandible. Be sure that all the parotid tissue has been removed at its posterior border. The masseter arises from the inferior surface of the zygomatic arch and maxilla and inserts on the lateral surface of the angle and ramus of the mandible. Review the relationships of the masseter to the parotid gland and duct and facial nerve. With the scalpel, carefully shave and free the branches of the facial nerve from the fascia on the lateral surface of the masseter, but do not detach them distally from the facial muscles. Also, free the parotid duct from this muscle. This maneuver will enable you to pull the masseter inferiorly and deep to the fibers of the facial nerve and parotid duct, as will be described.

To remove the zygomatic arch and attached fibers of the masseter, cut the arch posteriorly just anterior to the temporomandibular joint. Then, place a probe deep to the arch anteriorly and isolate it as far anteriorly as possible. Now, cut the zygomatic root of the arch along a line that passes through the bone at the anterior border of the masseter (Figure 9.17, A, Cut #2). Carefully pull the arch and attached fibers of the masseter inferiorly beneath the parotid duct and branches of the facial nerve, keeping the latter structures intact. The masseteric nerve (branch of the mandibular nerve) passes through the mandibular notch from the infratemporal fossa and enters the muscle on its deep surface. The masseteric nerve will be broken during reflection of the masseter. Continue to remove the fibers of the masseter from the lateral mandible, but leave them attached to the mandibular angle.

Step 3. The dissection now exposes the lateral surface of the man- dible and lower fibers of the insertion of the temporalis muscle. Note the vertical orientation of the fibers of the temporalis, which insert on the **coronoid process** inferiorly to the molar teeth. These fibers are surprisingly extensive and deep. Before removing them, identify the **buccal nerve** (pp. 76–78), a sensory branch of the mandibular nerve from the face and buccal mucosa of the oral cavity (pp. 76–78). The buccal nerve is closely embedded in and attached to the anterior, rounded tendon of the temporalis as it inserts posterior to the last molar tooth. The buccal nerve then becomes superficial at the buccal fat pad. Free the nerve from the insertion of the temporalis. With the buccal nerve identified, cut through the base of the coronoid process of the mandible and pull it and the attached fibers of the temporalis superiorly, thus reflecting the muscle from the temporal fossa (Figure 9.17, B, Cut #3) (p. 56). Place a probe posterior to the coronoid process to protect the deeper soft structures. Identify the **deep temporal vessels** and **nerves** (branches of the mandibular nerve) deep to the temporalis, where they lie against the bony floor of the temporal fossa (p. 82). Be sure to remove all the vertical fibers of the temporalis attached to the anterior border of the mandible as far as the last molar tooth that were not removed with the coronoid process. All the vertical fibers of the temporalis should be removed.

Step 4. Next, cut through the **neck of the mandible** just inferior to the temporomandibular joint (Figure 9.17, B, Cut #4). Then, carefully pull the ramus laterally to identify the **mandibular foramen** and **lingula** on the medial surface of the mandible, noting the **inferior alveolar nerve** and **vessels** that enter the foramen. With bone cutters, cut transversely through and remove the ramus of the mandible at a level just superior to the mandibular foramen (Figure 9.17, B, Cut #5). Be sure to protect your eyes from any flying bone fragments.

With these four procedures now completed, the lateral wall of the infratemporal fossa has been removed and its contents can now be dissected (Figure 9.17, C). Superficially in the fossa, locate the fragments of the **pterygoid plexus of veins** on the lateral surfaces of the

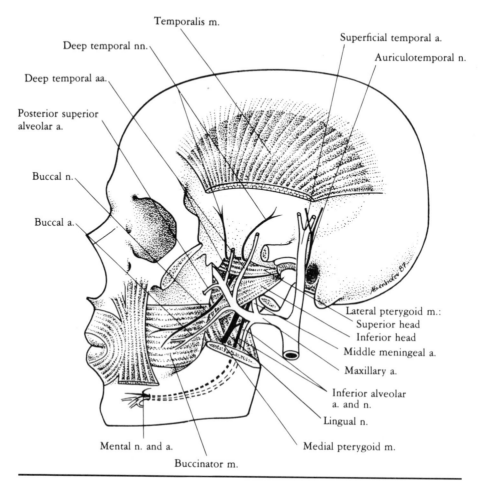

Figure 9.18

pterygoid muscles. Describe its tributaries and communications with other veins. Note that the pterygoid plexus forms the **maxillary vein,** which passes posterior to the neck of the mandible to join the **superficial temporal vein** to form the **retromandibular vein** in the parotid gland. Identify the **sphenomandibular ligament** attached to the lingula.

The **medial** and **lateral pterygoid** (p. 56) muscles should be identified next. Clean the fascia from these muscles and note their relationships to each other and directions of their fibers and describe their origins and insertions (Figure 9.18). The medial and lateral pterygoids are the last two muscles of mastication. The nerves to the pterygoids reach them on their deep surface but need not be identified in the laboratory. Review their actions. Identify the two heads of the lateral pterygoid and note their origins and relationships to the course of the **buccal nerve,** which passes between them. The insertions of the two heads will be dissected later. The maxillary artery may pass superficial or deep to the lateral pterygoid. Note this course in your cadaver.

Identify and follow the **lingual** and **inferior alveolar nerves** (pp. 78–79), which appear at the inferior border of the lateral pterygoid and cross the medial pterygoid laterally (Figure 9.18). These nerves are sensory branches of the mandibular nerve. Review their distributions and innervations. The **inferior alveolar nerve** enters the mandibular foramen and canal where it innervates the lower teeth before exiting as the cutaneous **mental nerve** at the mental foramen on the chin. Identify the **mylohyoid nerve,** which branches from the inferior alveolar nerve before the alveolar nerve enters the mandibular foramen. The mylohyoid nerve grooves the medial surface of the mandible to reach the submandibular region, where it innervates the mylohyoid and anterior digastric muscles. The **lingual nerve** passes between the mandible and the medial pterygoid to enter the submandibular region deep to the mylohyoid, where it has already been dissected.

Next, follow the course of the **maxillary artery** across the infratemporal fossa (pp. 76–82). The artery branches from the external carotid artery within the parotid gland and passes posterior to the neck of the mandible to enter the infratemporal fossa. In the fossa, the artery courses across the infratemporal fossa, superficial or deep to the lateral pterygoid, to enter the deeper pterygopalatine fossa via the pterygomaxillary fissure (Figure 9.18). Observe which course the artery takes in your cadaver. The first two branches of the maxillary artery are the deep auricular and anterior tympanic arteries to the external and middle ear. These branches are extremely small and often are difficult to locate in the laboratory. The first easily recognizable branch is the **inferior alveolar artery** (p. 76). This vessel descends superficial to the medial pterygoid and joins the **inferior alveolar nerve** to enter the mandibular foramen. Opposite the origin of the inferior alveolar artery, the middle meningeal artery branches from the maxillary artery and courses superiorly and deep to the lateral pterygoid (p. 81). The deep course of this artery will be described later. Other branches of the maxillary artery are the arteries to the four muscles of mastication and the **buccal artery,** which passes between the two heads of the lateral pterygoid with the **buccal nerve** (p. 76). Venous tributaries to the pterygoid plexus of veins accompany each of these arteries.

86. DEEP DISSECTION OF THE INFRATEMPORAL FOSSA

The remaining structures in this fossa are deep to the lateral ptery-goid and will be dissected after removal of this muscle on only one side (Figure 9.19). Identify the superior and inferior borders and origins of the two heads of the lateral pterygoid from the lateral surface of the **lateral pterygoid plate** and **infratemporal surface** of the **sphenoid bone.** Carefully cut these fibers and blunt dissect the muscle laterally toward its insertions to the mandibular condyle and articular disc of the temporomandibular joint. Clean away any loose tissue superficial to the joint and open the lateral surface of the joint capsule to identify the **articular disc.** At the insertion of the lateral pterygoid, identify the fibers attached to the neck (fovea) of the mandibular condyle (inferior head) and articular disc of the temporomandibular joint (superior head) (pp. 52–56). Remove the lateral pterygoid and attached structures (condyle and articular disc) of the temporomandibular joint. Identify and protect the course of the **maxillary artery** superficial or deep to the muscle.

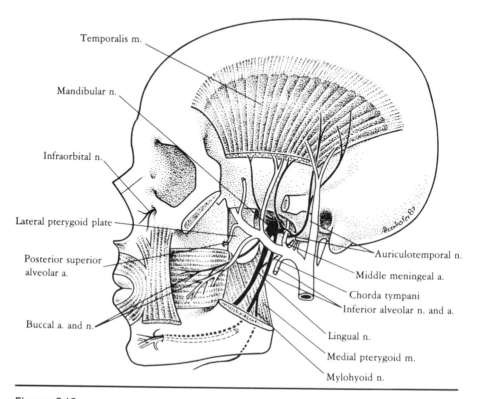

Figure 9.19

The temporomandibular joint is a synovial articulation between the **mandibular fossa** and the **condyle of the mandible.** Identify these bony landmarks on the skull. Review the movements of the joint and function of the masticatory muscles.

Deep to the lateral pterygoid, identify the course of the **middle meningeal artery,** trunk of the **mandibular nerve, lingual** and **inferior alveolar nerves, auriculotemporal nerve** (which encircles the middle meningeal artery), **chorda tympani nerve, deep temporal nerves** and **buccal nerve** (Figure 9.19) (pp. 80–82). Elevate the inferior alveolar nerve to observe the chorda tympani as it joins the lingual nerve. Review the named branches of the maxillary artery. Follow the maxillary artery as it passes between the two heads of the lateral pterygoid to enter the pterygopalatine fossa through the pterygomaxillary fissure. Observe the deep temporal nerves, which pass over the infratemporal crest deep to the upper head of the lateral pterygoid to enter the temporal fossa.

87. NASAL CAVITY

Use a dried skull and your textbook to define all the bony landmarks related to the roof, floor (hard palate) and lateral and medial walls of the nasal cavity (pp. 45–47). Note the foramina that open into the cavity. Also, name and identify the positions of the four large paranasal sinuses.

Examine the nasal septum, which was left intact on one side of the bisected head. This septum forms the medial wall of the nasal cavity. Peel the nasal mucosa from the septum and observe its bony and cartilaginous components (**vomer, perpendicular plate of the ethmoid bone** and **septal cartilage**) (p. 133). Although difficult to dissect, the small vessels and nerves (anterior ethmoidal vessels and nerves, nasopalatine nerve and sphenopalatine artery) that supply the septum are located in the dense mucosa. The bones of the hard palate (**palatine process of the maxilla** and **horizontal plate of the palatine bone**) form the floor of the nasal cavity. Identify these bones and note their relationships to the septum. The roof is formed primarily by the **cribriform plate of the ethmoid bone.** Note that this area is extremely narrow. The nasal (olfactory) mucosa of the roof contains terminals of the olfactory nerves, which pass through the cribriform plate to the anterior cranial fossa.

The lateral nasal wall (Figure 9.20) requires more detailed observations (p. 45). This bony wall is more complex than the medial wall and its major components should be identified (three **conchae, maxilla, vertical process of the palatine bone,** and **medial pterygoid plate**) by use of a dried skull and your textbook. The primary fea-

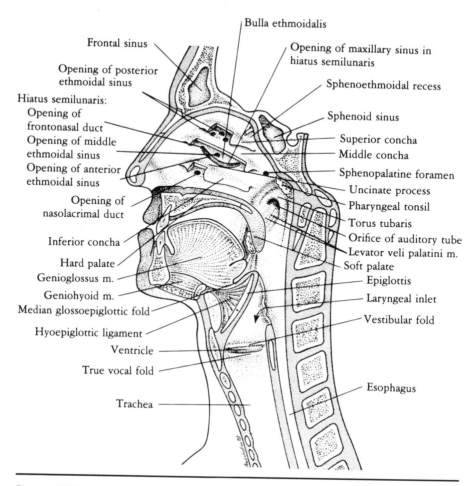

Figure 9.20

tures of the lateral wall are the **superior, middle** and **inferior nasal conchae** (pp. 84; 134), which project medially into the nasal cavity and almost touch the septum (Figure 9.20). The superior and middle conchae are parts of the ethmoid bone, but the inferior concha is a separate bone of the skull. In the thin mucosa that covers these conchae are small branches of the lateral nasal nerves and vessels. The spaces inferior to the conchae are the **superior, middle** and **inferior meatuses,** respectively. Identify these meatuses and note the position of the **sphenoethmoidal recess** superior to the superior concha. The relationships in the meatuses that will be described should be observed by elevating and removing the inferior and middle conchae on only one side (p. 134). The **nasolacrimal canal** and **duct** drain from the orbit into the inferior meatus deep to the inferior concha. Identify this canal on a dried skull. The opening of the canal is often difficult to identify in the cadaver. Deep to the middle concha, observe the **bulla ethmoidalis, hiatus semilunaris** and **uncinate process of the ethmoid bone** (Figure 9.20). The middle ethmoidal air sinus opens at the bulla and the maxillary, anterior ethmoidal and frontal (frontonasal duct) sinuses open at the hiatus semilunaris. Break through the lateral nasal wall and note the large **maxillary sinus** lateral to the nasal cavity. The posterior ethmoidal air sinuses open into the superior meatus. The complex labyrinths of the anterior, middle and posterior **ethmoidal air cells** are lateral to the middle and superior conchae. Note that the ethmoidal sinuses are between the nasal cavity medially and the orbit laterally. The **sphenoid sinus** drains into the sphenoethmoidal recess. Examine the large **sphenoid** and **frontal sinuses** and note that their partitions are often incomplete.

Strip the mucosa from the perpendicular (vertical) plate of the palatine bone posterior to the middle concha to identify the **sphenopalatine foramen.** This foramen connects the nasal cavity to the more lateral pterygopalatine fossa. Look for the **nasopalatine nerve** and **sphenopalatine artery,** which pass through the foramen to enter the nasal mucosa.

88. ORAL CAVITY

This area will be studied from the medial aspect of the bisected head (Figure 9.20). The oral cavity is divided into a vestibule and an oral cavity proper. The vestibule is the narrow area between the lips and cheeks and the teeth and gums. The parotid duct opens into the vestibule at the second upper molar tooth. The oral cavity proper is bounded by the teeth anteriorly and laterally and extends posteriorly to the oropharynx, with its aperture bounded by the **palatoglossal** and **palatopharyngeal arches.** Inferiorly, the tongue forms much of the floor of the cavity and the hard palate its roof. Identify these major landmarks. All the surfaces are covered by mucous membranes and innervated by branches of the maxillary and mandibular nerves, as described. The mucosa of the roof is innervated by the greater palatine nerve in the posterior two-thirds and by the nasopalatine nerve in the anterior third. The cheeks are innervated by the buccal nerve (derived from the mandibular nerve) and the mucosa of the floor by the lingual nerve. The palatal arches are innervated by the glossopharyngeal nerve. Review the orientation of dentition in the adult and define the major innervations to the maxillary and mandibular alveolar teeth and gums.

At the **floor** of the mouth, elevate the tongue and observe the **frenulum** of the tongue and openings of the two **submandibular ducts** on either side of the frenulum. The **sublingual folds,** elevated by the underlying sublingual glands, are also visible on the lateral aspects of the floor. Carefully cut and strip away the mucosa on the floor and review from above the following structures, which were dissected earlier with the submandibular region: **lingual nerve, deep part of the submandibular gland** and **duct, submandibular ganglion, sublingual gland, lingual artery, hypoglossal nerve** and **geniohyoid** and **mylohyoid** muscles (pp. 139; 140). Review these structures in the submandibular region.

The tongue should now be studied. At the lateral surface of the tongue in the submandibular region, identify the extrinsic muscles: the **styloglossus, hyoglossus** and **genioglossus** (pp. 137–138). All these muscles are innervated by the hypoglossal nerve. Define the origins and review the actions of these muscles on the tongue. Note their relationships to other structures in the floor of the mouth. In the submandibular region and floor of the mouth, follow the hypoglossal nerve laterally across the hyoglossus and genioglossus into the body of the tongue. This nerve innervates the extrinsic and intrinsic muscles of the tongue (note these muscles on the median surface of the cut tongue) (p. 137). Review the course of the lingual artery deep to the hyoglossus, where it supplies the tongue and floor of the mouth. Review the lingual nerve, which carries fibers of general sensation from the anterior two-thirds of the tongue. In addition, fibers of the chorda tympani nerve (VII) join the lingual nerve in the infratemporal fossa and supply taste buds on the anterior two-thirds of the tongue and parasympathetic, preganglionic secretomotor fibers to the submandibular and sublingual glands. These parasympathetic fibers synapse in the submandibular ganglion.

89. PALATE AND PHARYNGEAL CAVITY

The **hard** and **soft palates** separate the oral cavity inferiorly from the nasal cavity and nasopharynx superiorly (Figure 9.20). The hard palate is formed by the **palatine process of the maxilla** and **horizontal part of the palatine bones** (p. 43). It is covered on its oral surface by a dense mucous membrane that contains palatine glands. Identify in the hard palate the **incisive, greater palatine** and **lesser palatine foramina** on the dried skull (p. 43). These foramina conduct branches of the maxillary nerve and artery to the palate and oral cavity.

The soft palate is a mobile, muscular structure attached to the posterior border of the hard palate. The mucosa contains many palatine glands on its oral surface. Strip away the mucosa to expose the **musculus uvulae,** which forms the bulk of the uvula and projects posteriorly and downward in the midline (p. 137). Laterally, the soft palate is continuous with the pharyngeal wall, forming two mucosal folds and muscles. Identify the **palatoglossal fold** and **muscle** anteriorly and the **palatopharyngeal fold** and **muscle** posteriorly. These two folds separate the oral cavity from the pharynx. Remove the mucosa from these folds to observe these muscles. Between these folds is the tonsillar bed, which contains the **palatine tonsils** (p. 134). Blunt dissect deep to the tonsil, if present, to separate its capsule from the loose connective tissue bed. With the gland and fascia removed, the **superior pharyngeal constrictor** and parts of the **palatopharyngeus** muscles can be seen in the floor of the palatine tonsillar bed. The **styloglossus** muscle is lateral to the tonsillar bed. Inferior to the superior constrictor muscle in the depth of the tonsillar bed, identify the final course of the **glossopharyngeal nerve.**

Superior to the soft palate in the nasopharynx, note the position of the medial opening of the **autidory tube** (p. 84). The cartilage of this tube overlaps the opening to form the **torus tubarius** (Figure 9.20). Observe the **salpingopharyngeal fold** (pp. 133; 134) inferior to the torus. Posterior to the salpingopharyngeal fold is the **pharyngeal recess.** Remove the mucosa from the lateral nasopharyngeal wall and floor of the auditory tube. Identify the **levator veli palatini** muscle as the large mass that elevates the mucosa on the floor at the medial opening of the auditory tube. The fibers of this muscle insert directly into the soft palate and elevate it. Anterior and more lateral to the levator veli palatini is the **tensor veli palatini.** To identify this muscle, expose the posterior edge of the medial pterygoid plate deep to the mucosa anterior to the auditory tube. Note the vertical fibers of the tensor veli palatini just lateral and posterior to the medial plate. The tensor originates from the scaphoid fossa at the base of the medial pterygoid plate. The fibers of this muscle descend vertically to the lateral side of the medial pterygoid plate and make a 90-degree turn around the hamulus of the medial pterygoid plate to insert into the aponeurosis of the soft palate. Dissect the course of the fibers of the tensor veli palatini around the hamulus to insert in the soft palate.

All the muscles of the palate are innervated by the pharyngeal branches of the vagus nerve, except for the tensor veli palatini, which is innervated by the mandibular nerve. The glossopharyngeal nerve supplies sensory fibers to most of the mucosa of the pharynx. Review the actions and attachments of the associated muscles of the palate.

90. PTERYGOPALATINE FOSSA

This fossa is small and located deeply in the head. It is wedged between the infratemporal fossa laterally and the nasal cavity medially (p. 44). In the pterygopalatine fossa, the distributions of the branches of the maxillary nerve and terminal part of the maxillary artery can be studied. Review the bony landmarks, boundaries and foramina related to this fossa from your textbook and identify these structures on the dried skull. Note on the skull the **greater** and **lesser palatine foramina,** which open into the oral cavity, and the **sphenopalatine foramen,** which connects to the nasal cavity. The **pterygomaxillary fissure,** through which the maxillary artery passes, connects laterally to the infratemporal fossa. The **inferior orbital fissure** connects to the orbit anteriorly and superiorly and the **pterygoid** and **pharyngeal canals** connect to the base of the skull posteriorly. Locate the pterygoid canal at the base of the medial pterygoid plate. The maxillary nerve exits the middle cranial fossa through the **foramen rotundum** and enters the pterygopalatine fossa.

The pterygopalatine fossa (Figure 9.21) will be dissected from the medial aspect of the bisected head. It is advisable to do this dissection on only one side of the head. Use the same side from which the conchae were removed in the nasal cavity. The sphenopalatine foramen was previously observed on the lateral nasal wall posterior to the middle concha. Remove the mucosa from the part of the lateral nasal wall inferior to the foramen to expose the **vertical plate of the palatine bone.** This bone is the medial wall of the pterygopalatine fossa. Carefully pick away the bone to expose the **greater** and **lesser palatine canals,** which contains the **greater** and **lesser palatine nerves** (Figure 9.21) and **descending palatine vessels** (p. 135). Follow these nerves superiorly and remove more bone to locate the **pterygopalatine ganglion** anterior to the sphenoid bone and **pterygoid canal** (p. 135). The **nasopalatine nerve** branches from the ganglion and passes through the sphenopalatine foramen to enter the nasal cavity. Only the stump of the nasopalatine nerve may be visible. Identify the **nerve of the pterygoid canal,** which exits the pterygoid canal and joins the ganglion just anterior to the floor of the sphenoid sinus (Figure 9.21). Note that the pterygoid canal elevates the floor of the sphenoid air sinus. Medial to the ganglion and superiorly in the fossa, locate the trunk of the **maxillary nerve.** Follow the maxillary nerve toward the floor of the orbit by removing a small part of the bones at the posterior orbital floor and posterior maxillary sinus. The continuation of the maxillary nerve on the orbital floor is the **infraorbital nerve.** This nerve passes through the

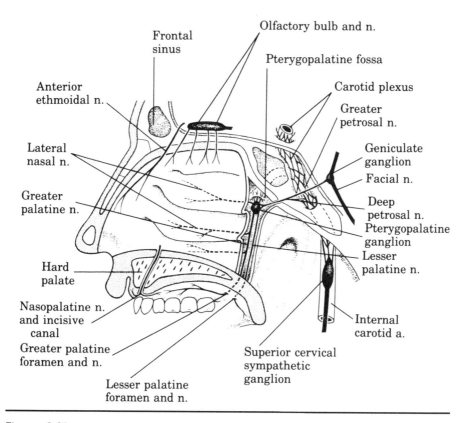

Figure 9.21

thin, bony roof (infraorbital groove) of the maxillary sinus and then
through the **infraorbital canal** to reach the face at the infraorbital
foramen, where it is cutaneous on the face (pp. 68; 70). Identify the
infraorbital nerve on the face deep to the levator labii superioris
muscle. The **anterior, middle** and **posterior superior alveolar
nerves** pass through the thin bone of the maxillary sinus.

Review the branches of the last part of the **maxillary artery,**
which distribute with the maxillary nerve (pp. 81–82). These arteries supply the palate, nasal cavity, maxillary teeth and face. In
the infratemporal fossa, follow the course of the middle part of the
maxillary artery as it enters the pterygopalatine fossa through the
pterygomaxillary fissure. From the medial and lateral aspects of
the pterygopalatine fossa, identify the **posterior superior alveolar,
infraorbital, descending palatine** and **sphenopalatine arteries.**
It may not be possible to find all these arteries on any one cadaver.
Review these branches and their distributions in the head.

91. ORBITAL REGION

The orbits are pyramid-shaped sockets that contain the eyeball and related structures. Each orbit has four walls, an apex at the optic canal and a base formed by the anterior margins of the orbit. Before dissecting the orbit, make a detailed study of these bony features and landmarks on the dried skull by use of descriptions from your textbook and atlas (pp. 45; 122). Note the relationships that will now be described. Superiorly, the roof of the orbit separates it from the anterior cranial fossa; medially, the orbit is related to the ethmoidal air sinuses; inferiorly, the floor of the orbit is the roof of the maxillary sinus; and laterally, the orbit borders the temporal fossa. Identify the **optic canal, anterior** and **posterior ethmoidal foramina** and **superior** and **inferior orbital fissures.**

A. Eyelid and Lacrimal Apparatus

Before dissecting the deeper structure within the orbit, a brief study of the eyelids and lacrimal apparatus should be made anteriorly from the face (p. 123). Review the orientation of the fibers of the **orbicularis oculi** muscle, noting its smaller palpebral part (pp. 58–59). Reflect the thin skin and muscle from the eyelids and identify the **tarsal plates** at the eyelid margin and the **orbital septum** (p. 123). The orbital septum is a fascial layer that attaches the tarsal plates to the bony anterior margin of the orbits, thus separating the orbital space from the surface of the face. The fibers of the **levator palpebrae superioris** muscle pierce the upper septum to insert on the fascia of the eyelid and the superior tarsal plate. At the medial and lateral angles of the orbit, the septum and tarsal plates attach to the medial and lateral palpebral ligaments. These ligaments attach to the frontal process of the maxilla and the zygomatic bones, respectively. Next, open the palpebral fissure and review the extent of the conjunctival sac. The conjunctiva is continuous from the deep surface of the eyelids onto the anterior surfaces of the sclera and cornea. Although you need not dissect them on the cadaver, understand the components of the lacrimal apparatus and the route of tear drainage from the surface of the eye to the nasal cavity. Note the position of the lacrimal gland in the superolateral margin of the orbit. Medially, the lacrimal papillae and puncta should be viewed in the living specimen. Tears are drained into the inferior meatus of the nasal cavity via the nasolacrimal duct. Note the **nasolacrimal canal** on the skull.

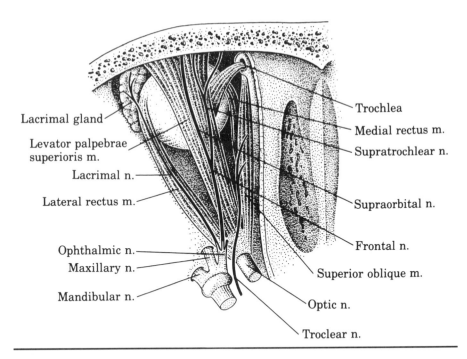

Lacrimal gland

Levator palpebrae
superioris m.

Lacrimal n.

Lateral rectus m.

Ophthalmic n.

Maxillary n.

Mandibular n.

Trochlea

Medial rectus m.

Supratrochlear n.

Supraorbital n.

Frontal n.

Superior oblique m.

Optic n.

Troclear n.

Figure 9.22

B. Orbit

The bony margins, walls and foramina related to the orbit were mentioned previously. Understand the arrangement of the **periorbita,
bulbar sheath** and **muscular fascia,** which line and cover the orbit
and its contents (p. 122). Review and understand the continuation of
the dura mater, arachnoid, pia mater and subarachnoid space along
the optic nerve from the cranial cavity to the eyeball.

The contents of the orbit will be dissected from above by removal of
its roof. In the floor of the anterior cranial fossa, strip away any
remaining dura and identify the **orbital plate of the frontal bone.**
This thin, flat plate of bone is the roof of the orbit. Break through the
bone and remove the entire orbital plate to expose the contents of the
orbit. Anteriorly and medially, you will encounter parts of the **frontal** and **ethmoidal air sinuses,** respectively. Posteriorly, a portion of
the **lesser wing of the sphenoid bone** at the anterior clinoid processes can be snipped away to expose the **superior orbital fissure
and optic canal.**

Initially, note the **periorbita** that encloses the orbital structures
as a sac. It is a dense, fascial covering but is easily separated from the
bony walls. Open the periorbita with an anteroposterior incision.

Most superficially in the orbit, identify medially at the apex the small **trochlear nerve** (Figure 9.22). In the cranial cavity, this nerve passes anteriorly in the lateral wall of the cavernous sinus and crosses superior to the oculomotor nerve to become the most superior nerve that enters the orbit. Identify the trochlear nerve as it crosses the frontal nerve to reach the **superior oblique** muscle (pp. 126–127). Next, follow the course of the **ophthalmic nerve** from the lateral wall of the cavernous sinus into the superior orbital fissure, where it divides into three main branches (pp. 130–131). Identify its two superficial branches, the **frontal** and **lacrimal nerves** (Figure 9.22). The frontal nerve passes in the midline toward the anterior orbital margin on the **levator palpebrae superioris** muscle, where it divides distally into the **supratrochlear** and **supraorbital nerves** (p. 131). These two nerves pass onto the skin of the forehead and scalp. The **lacrimal nerve** passes laterally at the superior border of the lateral rectus muscle to reach the **lacrimal gland** and adjacent eyelid and conjunctiva. The deeper and third nasociliary branch of the ophthalmic nerve will be dissected later.

Identify the seven extraocular muscles. Six of these muscles move the eyeball and the remaining one raises the upper eyelid (p. 126). The four **rectus** muscles (superior, inferior, lateral and medial) arise from the common ring tendon that encircles the optic canal and part of the superior orbital fissure. Superiorly, in the midline of the orbit, identify the **levator palpebrae superioris,** which overlies the deeper **superior rectus.** On the superomedial margin of the orbit, note the **superior oblique** muscle and follow its tendon around the pulley (trochlea) to its insertion on the eyeball. Observe the **medial** and **lateral recti** at their respective sides of the orbit. The **inferior rectus** is on the floor of the orbit and can be identified by viewing it from the dissection of the pterygopalatine fossa. The **inferior oblique** is demonstrated anteriorly on the face by cutting the orbital septum of the lower eyelid and elevating the eyeball (p. 126). Understand the insertions of these muscles on the surface of the eyeball and review their actions and innervations.

Cut and reflect the levator palpebrae superioris and superior rectus superiorly in the orbit at their midpoints (Figure 9.23) to dissect the deeper structures of the orbit (p. 131). It is now possible to identify the third and deepest branch of the ophthalmic nerve, the **nasociliary nerve.** Follow this nerve medially as it passes the **optic nerve** deep to the superior rectus to reach the medial wall of the orbit. Identify the **long ciliary, anterior ethmoidal** and **posterior ethmoidal branches** of the nasociliary nerve (Figure 9.23). At the lateral side of the orbit is the **abducens nerve.** This nerve enters

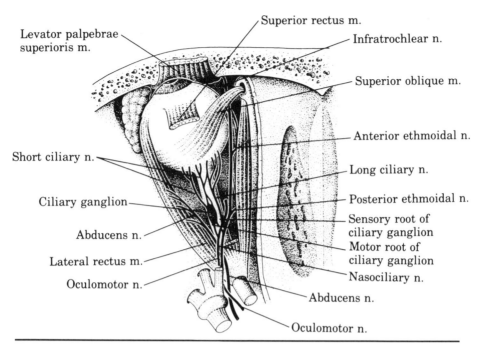

Levator palpebrae superioris m.

Superior rectus m.

Infratrochlear n.

Superior oblique m.

Anterior ethmoidal n.

Short ciliary n.

Long ciliary n.

Ciliary ganglion

Posterior ethmoidal n.

Sensory root of ciliary ganglion

Abducens n.

Motor root of ciliary ganglion

Lateral rectus m.

Nasociliary n.

Oculomotor n.

Abducens n.

Oculomotor n.

Figure 9.23

the orbit through the superior orbital fissure and can be identified on, or embedded in, the medial surface of the lateral rectus.

The **oculomotor nerve** enters the orbit through the superior orbital fissure as it passes between the two heads of the lateral rectus. Identify its **superior** and **inferior branches** and review their innervations and distributions. The superior branch to the superior muscles passes medially across the optic nerve with the nasociliary branch of the ophthalmic nerve, whereas the inferior branch courses forward, deeper in the orbit. In addition to its motor innervations, the oculomotor nerve carries parasympathetic preganglionic fibers, which synapse in the ciliary ganglion. The postganglionic fibers then run in the short ciliary nerves to the globe to innervate the sphincter pupillae muscle of the iris and the ciliary muscle of accommodation. Locate the small **ciliary ganglion** (p. 131), which is wedged between the optic nerve and the lateral rectus at the apex of the orbit (Figure 9.23). This parasympathetic motor ganglion is functionally related to the oculomotor nerve. A **motor root** from the inferior branch of the oculomotor nerve joins this ganglion to carry preganglionic parasympathetic fibers from the oculomotor nerve to the ciliary ganglion, where they synapse, as was discussed previously. The ciliary ganglion is also connected by a **sensory root** to the nasociliary nerve, although the fibers of this nerve do not synapse in the ganglion. A sympathetic motor root carries postganglionic fibers that pass through the superior cervical ganglion without synapsing to innervate the dilator pupillae muscle. These autonomic and sensory fibers are distributed to the globe via the short ciliary nerves. Identify the **short ciliary nerves** between the ganglion and the globe.

Follow into the orbit the **ophthalmic artery** (p. 125) after it branches from the internal carotid artery. The artery crosses the optic nerve and runs with the nasociliary nerve. In the orbital cavity, many of its named branches correspond to branches of the ophthalmic nerve. Find examples of as many of these arterial branches as possible. Also, identify the **ophthalmic veins** (superior and inferior) and review their communications with the veins on the face, cavernous sinus and pterygoid plexus of veins.

92. EAR

This dissection will provide a brief review of the major structures and relationships of the ear. As you proceed, carefully compare your work with your atlas and textbook. Initially, make a study of the petrous portion of the temporal bone, which contains the middle and internal ear. Project the positions of the structures to the surface of the petrous bone as they are studied.

A. Inner Ear

Note the **facial** and **vestibulocochlear nerves** as they enter the **internal acoustic meatus** (pp. 120–121). Use the skull and your textbook to review the course of the facial canal through the petrous part of the temporal bone. The orientation of the facial canal is important in understanding the positions of structures in the petrous bone. The inner and middle ear is best dissected by the technique described by Cahill and Snow.* First, make two parasagittal saw cuts through the petrous part of the temporal bone, one just lateral to the internal auditory meatus and the other at the lateral end of the petrous bone where it joins the cranial wall (Figure 9.24, Cut #1). These saw cuts should be extended deeply to the plane of the floor of the middle cranial fossa anteriorly and to the level of the internal acoustic meatus posteriorly. The internal structures of the petrous bone are exposed by chiseling against the posterior surface of the

*Cahill DR, Snow MH: A quick, effective method for dissecting the middle ear. *Anat Rec* 181:685–688, 1975.

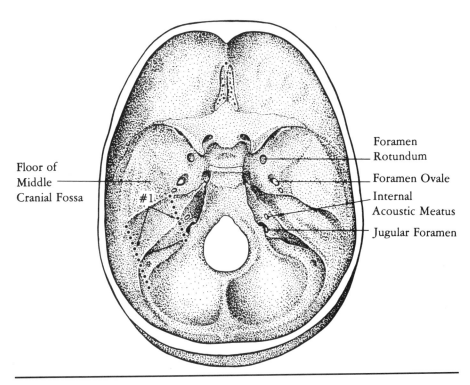

Foramen
Rotundum

Foramen Ovale

Internal
Acoustic Meatus

Jugular Foramen

Floor of
Middle
Cranial Fossa

#1

Figure 9.24

temporal bone and breaking away the **tegmen tympani** and upper surface of the petrous bone. Note that the chisel should be held parallel to the floor of the middle cranial fossa. With the bone removed, identify the **greater petrosal nerve** and **facial nerve** with its external genu and **geniculate ganglion**. Carefully chip away the petrous bone anterior to the facial and vestibulocochlear nerves to demonstrate the **cochlea** (p. 120) and observe the orientation of the cochlea in the bone. Next, chip away the bone posterior to the facial canal to expose one or more of the semicircular canals and the vestibule (p. 120).

B. Middle Ear

The middle ear, or tympanic cavity, is the small, central division of the ear. The tegmen tympani of the petrous bone forms the roof of the middle ear and was chipped away previously. Lateral to medial, the hour-glass-shaped tympanic cavity (pp. 118–119) is narrow (2 to 6 mm). The main relationships of the middle ear are as follows: anteriorly, the auditory tube and carotid canal; laterally, the tympanic membrane and chorda tympani nerve; posteriorly, the mastoid air cells; medially, the inner ear; superiorly, the middle cranial fossa; and inferiorly, the jugular fossa.

With the tegmen removed, identify the **bony ossicles** that cross lateral to medial across the space. The **malleus** and **incus** are more easily seen than is the **stapes.** At the lateral wall, find the **chorda tympani nerve** (pp. 118–119), which passes medial to the handle of the malleus and the tympanic membrane. Most laterally, identify the **tympanic membrane.** Attempt to locate the tendon of the **tensor tympani** muscle (innervated by the mandibular nerve), which inserts on the malleus. This muscle passes from the roof of the auditory tube at the anterior wall. Posteriorly are the **pyramid** and projecting tendon of the **stapedius** muscle, which inserts on the stapes.

Although not totally visible in the dissection, review the major structures on the medial wall (promontory, fenestra vestibuli, fenestra cochleae, prominence of the facial canal and prominence of the lateral semicircular canal). The course of the facial canal should be reviewed and its relationships to the medial and posterior walls of the middle ear should be noted. The mucosa of the middle ear is innervated by the tympanic branch of the glossopharyngeal nerve, which penetrates the floor of the tympanic cavity.

C. External Ear

The external division of the ear is formed by the auricle and external acoustic meatus (p. 116). Review the course and direction of the external meatus (bony and cartilaginous parts). Note the position and structure of the tympanic membrane at the medial end of the external meatus. Review the innervations of the auricle and external acoustic meatus (branches of the facial, trigeminal, glossopharyngeal and vagus nerves).